AN INTRODUCTION TO
Peptide Chemistry

To Judith

I'd like to acknowledge the invaluable advice, constructive criticism, and encouragement not only from numerous university and industrial colleagues, but also from the undergraduate and graduate students at York.

AN INTRODUCTION TO
Peptide Chemistry

P. D. BAILEY
University of York

JOHN WILEY & SONS

Chichester · New York · Brisbane · Toronto · Singapore

SALLE + SAUERLÄNDER

Aarau · Frankfurt am Main · Salzburg

Copyright © 1990 by John Wiley & Sons Ltd, Baffins Lane,
Chichester, West Sussex PO19 IUD, England
Otto Salle Verlag GmbH & Co., Frankfurt am Main
Verlag Sauerländer AG, Aarau

Reprinted with corrections April 1992
Reprinted June 1997

Other Wiley Editorial Offices

John Wiley & Sons, Inc., 605 Third Avenue,
New York, NY 10158-0012, USA

Jacaranda Wiley Ltd, G.P.O. Box 859, Brisbane,
Queensland 4001, Australia

John Wiley & Sons (Canada) Ltd, 22 Worcester Road,
Rexdale, Ontario M9W 1L1, Canada

John Wiley & Sons (SEA) Pte Ltd, 37 Jalan Pemimpin   05-04,
Block B, Union Industrial Building, Singapore 2057

Library of Congress Cataloging-in-Publication Data:
Bailey, P. D.
      An introduction to peptide chemistry   /   by P. D. Bailey.
            p.   cm.
      Includes bibliographical references.
      ISBN 0 471 92348
      1. Peptides.   I. Title.
      QD431·B235   1990   89-22582
      547.7′56—dc20   CIP

British Library Cataloguing in Publication Data:
Bailey, P. D.
      An introduction to peptide chemistry.
      1. Peptides
      I. Title
      547.7′56

      ISBN 0 471 92348 6 ppc
      ISBN 0 471 93532 8 pbk

CIP-Titelaufnahme der Deutschen Bibliothek:
Bailey, Patrick D.:
An introduction to peptide chemistry   /   by P. D. Bailey.—
Chichester ; New York ; Brisbane ; Toronto : Wiley ; Aarau ;
Frankfurt am Main ; Salzburg : Salle u. Sauerländer, 1990
      ISBN 3-7935-5546-1 (Salle u. Sauerländer) Gb. ppc
      ISBN 3-7935-5544-5 pbk
      ISBN 0-471-92348-6 (Wiley) Gb. ppc
      ISBN 0 471 93532 8 pbk

Typeset by Thomson Press (I) Ltd. New Delhi
Printed and bound in Great Britain by Bookcraft (Bath) Ltd, Midsomer Norton, Somerset

Contents

CHAPTER 1

Introduction

The peptides are an amazing class of compounds! Although they are all constructed from relatively simple building blocks (the amino acids), they exhibit a remarkable range of biological properties: peptides can act as antibiotics, hormones, food additives, poisons, or pain-killers. And it is primarily because of their medicinal properties that the study of peptides has become one of the most active areas of current research. In this book, we will look at the basic chemistry which governs how peptides can be isolated, how their structure can be determined, and how they can be synthesised.

1 What are Peptides?

When amino acids are covalently linked together by amide bonds, the resulting molecules are called **peptides** or **proteins**.

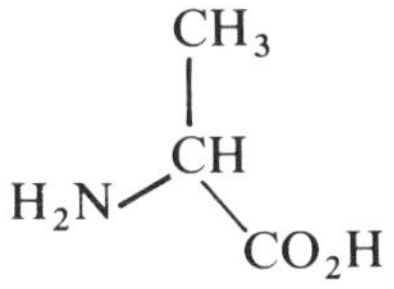

Alanine; an amino acid

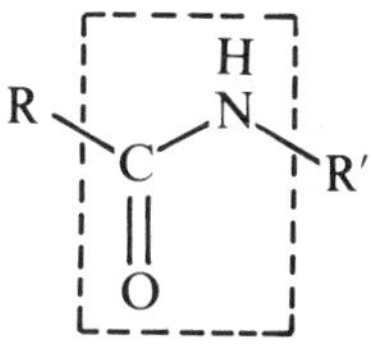

The amide bond

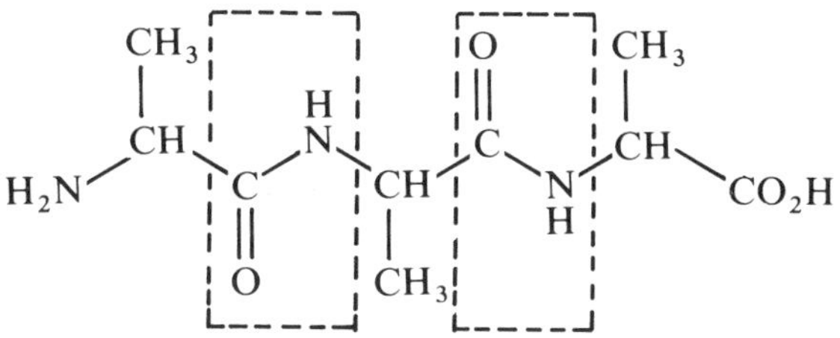

A tripeptide, composed of three alanine residues

> The **amide bond**
> that links together
> the **amino acids** in
> **peptides and proteins** is
> called the **peptide bond**.
>
> **Amino acids** that
> are part of a
> **peptide or protein**
> are referred to
> as **residues**.

Although both peptides and proteins are composed of amino acid residues, there is a subtle difference between these two types of molecule.

Proteins are large molecules that usually contain at least 50 residues—and sometimes over 1000. The most important feature of proteins is that they possess well-defined three-dimensional structures.

For example, haemoglobin, which transports oxygen from the lungs to the rest of the body, has a precise shape which allows oxygen to bind reversibly. And carboxypeptidase A is only one of many enzymic proteins that break down food in the stomach; each such protein has a specific structure that allows the catalytic hydrolysis of one particular type of bond. In contrast, skin has a much less precise molecular structure, but its toughness is mainly due to a protein called collagen which is composed of well-defined helical rods; these can then form inter-linking strands rather like a woven fabric. So whatever the function of a particular protein, its properties depend upon its precise three-dimensional conformation.

> For peptides and proteins, the **primary** structure is just the order (or sequence) of the amino acid residues. **Secondary** structure refers to specific structural features within the molecule. **Tertiary** structure is the overall three-dimensional shape of the molecule.

Peptides, on the other hand, are rather smaller molecules (usually containing less than 50 residues), which do not generally possess a well-defined three-dimensional structure. The distinction between peptides and proteins is sometimes rather unclear, particularly with medium-sized molecules of 15 to 50 residues; these latter compounds are often referred to as polypeptides.

1.1 The Relative Size of Peptides

The approximate sizes of a range of molecules are indicated in Figure 1.1, including certain amino acids, peptides, and proteins. The representations are only intended to give a rough idea of the relative volumes occupied by these molecules, and the peptides are likely to have rather open, 'floppy' structures in solution. In contrast, proteins generally have very compact structures, with well-defined three-dimensional shapes (tertiary structure); haemoglobin and tropocollagen

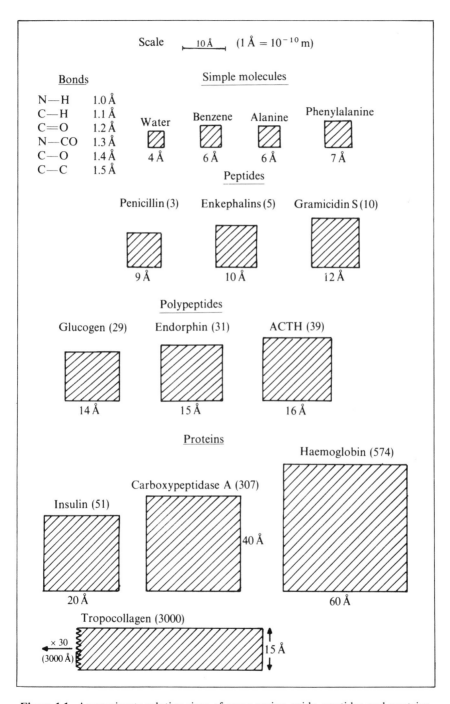

Figure 1.1. Approximate relative sizes of some amino acids, peptides and proteins.

(the building block of collagen) are in fact composed of separate protein units that lock together, and this is known as quaternary structure. All of the molecules shown have important biological properties (e.g. the polypeptides are all vital human hormones), and the numbers in brackets refer to the number of constituent amino acids.

There is one other feature that is common to all proteins: their biosynthesis is directed by the genetic code contained in DNA, and this means that they are composed only of the 20 common amino acids. Many peptides are also DNA encoded, but there are many that are made by different pathways, and that contain unusual amino acids.

Because of the size, structure, and genetic origin of proteins, a range of special techniques have been developed for their study. Rather different approaches are often used for research into peptides, although there is obviously quite a lot of overlap. So, although this book is specifically about peptides, there will be frequent comparisons with the techniques that are used for studying proteins.

2   Properties of the Peptide Bond

Although peptides are composed of amino acids, the amide bond itself shows neither the properties of the amino group, nor those of the carboxylic acid group. In fact, the properties of the amide group are governed by the **conjugation** of the nitrogen lone pair with the carbonyl group—this **mesomeric** effect can be expressed as a **resonance** between two **canonical** forms.

The right-hand canonical form contributes significantly to the properties of the amide bond, even though it involves charge splitting; not only do all the atoms possess complete outer shells of electrons, but the charge distribution is also favourable (cf. RO^- and RNH_3^+ are relatively stable ions).

There are three important consequences of the resonance stabilisation of the amide group.

(i) The peptide bond is rather inert chemically, because any reactions would disrupt the conjugation. For example, the nitrogen atom is neither basic nor nucleophilic, because the lone pair of electrons is conjugated with the carbonyl group (cf. the properties of amines). Conversely, the carbonyl group can be attacked by nucleophiles, but forcing conditions are usually required because the carbonyl group is stabilised by interaction with the nitrogen lone pair.

(ii) The amide group can occasionally act as a nucleophile—in which case it is usually the oxygen (which carries a partial negative charge) which is the nucleophilic atom, rather than the nitrogen.

(iii) The amide group is planar. This is required in order that the nitrogen lone pair can interact with the carbonyl π-bond; i.e. the nitrogen, carbon, and oxygen p-orbitals all have to lie in the same plane, so that conjugation can take place.

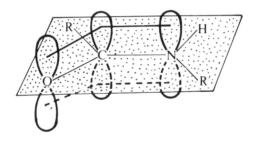

The planarity of the amide bond is particularly important to the conformation (or three-dimensional shape) of peptides.

3   What Biological Properties do Peptides Possess?

The reason that special interest is shown in peptides is that many of them possess potent pharmacological properties. This means, for example, that they have potential use as medicines, provided that they can be isolated and synthesised. To give a flavour of the wide range of biological properties of peptides, a few examples are given in Table 1.1.

Table 1.1. Biological properties of some amino acids and peptides.

Name	No. of residues	Biological property
GABA	1	Neurotransmitter, involved in the control of nerve impulses
Monosodium glutamate	1	A 'meaty' flavoured food additive
Aspartame	2	An artificial sweetener, about 100 times sweeter than sucrose
Penicillin	3	A powerful antibiotic, which is formed in certain fungal moulds
TRH	3	A hormone that controls the release of another hormone (thyrotrophin) in the body, and also affects the central nervous system
Enkephalins	5	Found in the brain, these peptides are involved in the control of the sensation of pain
Phalloin	7	An extremely poisonous bicyclic peptide, found in the Deathcap toadstools
Angiotensin II	8	Used by the body to increase blood pressure, this peptide is known as a hypertensive agent
Oxytocin	9	A hormonal cyclic nonapeptide which, among other things, can be used to induce labour in pregnancy
Gramicidin S	10	A cyclic decapeptide that is a powerful antibiotic

So, from a biological point of view, peptides can act as hormones, poisons, antibiotics, ..., and show a diverse range of other properties.

The medicinal applications are not hard to see (but they are sometimes hard to put into practice). For example, angiotensin II (an octapeptide) is produced by our bodies, and causes an increase in blood pressure; so synthetic angiotensin II, or peptidic drugs that could **mimic** its action, might be used in the treatment of **low** blood pressure, whilst peptides that could **block** its action might be used to alleviate **high** blood pressure. On the other hand, the naturally occurring antibiotic penicillin (which is a modified tripeptide) is known to destroy many bacterial infections, but chemists are continually synthesising new derivatives that will combat resistant bacteria, or that have fewer side-effects with patients.

There is insufficient space in this book for us to look in detail at the medicinal applications of peptide chemistry, but it is certainly possible that synthetic peptides could dominate the pharmaceutical market in the near future. It is also worth adding that effective drugs are of considerable financial value.

4 How are Peptides Studied?

Suppose a new peptide were discovered, and that it triggered an important biological response; what experiments would we need to carry out, in order to prepare a drug that might have the same function?

(i) We would have to isolate the peptide, and then ensure that we had obtained the correct molecule in a really pure state.
(ii) We would need to determine the nature and quantity of the constituent amino acids present in the peptide.
(iii) Next, the actual primary structure of the peptide itself would have to be determined (i.e. the order, or sequence, of the constituent amino acids).
(iv) Finally, we would have to synthesise the peptide; this would enable us to confirm that we had determined its structure correctly, and might also allow us to prepare a drug for medicinal purposes.

For the development of a drug with modified biological properties (perhaps greater activity, or fewer side-effects), we might try to find out more about the detailed structure of the peptide, and how it interacts with other molecules in the body. This could help us to design new molecules, which would need to be synthesised and tested. This is the rather more advanced topic of medicinal chemistry, and is beyond the scope of this book. However, the four stages outlined above are absolutely crucial to any studies on naturally occurring peptides.

5 The Layout of this Book

In order to carry out the four stages above, it is vital that we should know as much as possible about the physical and chemical properties of the constituent amino acids. This book is therefore divided into five main chapters:

Chapter 2. This looks at the properties of the **amino acids** that are present in peptides.
Chapter 3. This considers the **isolation and purification** of peptides.
Chapter 4. Amino acid analysis allows the constituent amino acids to be identified and quantified (but their order would still be unknown).
Chapter 5. The **sequencing** of peptides is discussed next, which allows the primary structure of peptidic molecules to be determined.
Chapter 6. This reveals how the **synthesis** of peptides might be accomplished.

Throughout this book, we will **imagine** that a new peptide has been found to be present in our bodies, which we will call **PENTIN**. This peptide can be found in the cartilage of healthy joints, but its levels are low in arthritic joints. In arthritis, the cartilage (and surrounding bone) are broken down by the action of certain proteases (enzymes that digest proteins); PENTIN might inhibit the action of these proteases, and so could be of immense medicinal (and commercial) value for the treatment of arthritis. In the following few chapters, we will isolate, sequence, and finally synthesise PENTIN, in order to be able to study its medicinal properties.

PENTIN is an imaginary peptide, whose isolation, structure determination, and synthesis will guide us through much of the peptide chemistry in this book.

In **Chapter 7**, we will see how the isolation, sequencing, and synthesis of a vital human hormone called LH-RH was actually accomplished in the 1970s. The molecule was a particularly difficult peptide to study because it could only be isolated in minute quantities, and the chief scientists were awarded the Nobel Prize for Medicine in 1976.

There are three appendices at the back of this book:

Appendix A looks at how one could go about synthesising amino acids themselves. This is not usually important for peptides that contain **only** DNA encoded residues (the constituent amino acids being readily available from natural sources), but new laboratory-made drugs often incorporate unusual amino acids within a normal peptide sequence; furthermore, many important naturally occurring peptides (e.g. several major antibiotics) contain amino acids that are not DNA encoded.

Appendix B looks at some of the ways in which peptides can fold up, and how the structure can be studied. Many peptides are known to have **some** well-defined three-dimensional features (secondary structure), and a number of cyclic peptides have very precisely known structures (tertiary structure); however, for peptides of less than 10 residues, clear structural features are rare, and so this topic is covered only briefly.

Appendix C gives an overview of some of the aspects of DNA/RNA technology that can be used for the sequencing or synthesis of peptides. In fact, the techniques of genetic engineering are particularly suitable for research into the structure and function of proteins; but the techniques have also been successfully applied to the study of a number of peptides, and they will undoubtedly become more important as further advances are made in this area of molecular biology.

Throughout this book, certain statements are enclosed in boxes; these are particularly important new concepts or definitions, that will be referred to later on in the text. At the end of most of the chapters, the chemistry of the preceding pages is applied to our imaginary peptide PENTIN; this essentially gives a summary of each aspect of peptide chemistry. Sources are also given for more advanced reading, so that you can delve a little more deeply into the specific topics that particularly interest you. Finally, the questions at the end of each chapter allow you to check that you have really understood the chemistry involved in each section.

The chemistry of amino acids and peptides is, for the most part, fairly straightforward (believe it or not). But the particularly satisfying feature of work involving them is that it brings together ideas from organic, physical, analytical, and biological chemistry.

Further Reading

There are dozens of books on different aspects of peptide chemistry, and an enormous number of papers are published in the journals. Below are given some selected general texts and sources, which you might find particularly helpful and/or interesting. Further reading on specific areas of peptide chemistry are given at the end of the appropriate chapters.

Multi-volume Texts

The Peptides; Analysis, Synthesis, Biology, Volumes 1–5 (E. Gross and J. Meienhofer, Eds.) and Volumes 6–9 (S. Udenfriend and J. Meienhofer, Eds.), Academic Press, 1979–87. An excellent series, that looks in some detail at many aspects of peptide chemistry and biochemistry:

Vol. 1—*Major Methods of Peptide Bond Formation*
Vol. 2—*Special Methods in Peptide Synthesis*, Part A
Vol. 3—*Protection of Functional Groups in Peptide Synthesis*
Vol. 4—*Modern Techniques of Conformational, Structural, and Configurational Analysis*
Vol. 5—*Special Methods in Peptide Synthesis*, Part B
Vol. 6—*Opioid Peptides: Biology, Chemistry, and Genetics*
Vol. 7—*Conformation in Biology and Drug Design*
Vol. 8—*Chemistry, Biology, and Medicine of Neurohypophyseal Hormones and Their Analogs*
Vol. 9—*Special Methods in Peptide Synthesis*, Part C

Chemistry and Biochemistry of Amino Acids, Peptides, and Proteins, Volumes 1–7 (B. Weinstein, Ed.), Dekker, 1971–83. A survey of 'recent' developments up to 1983; each volume covers six or so topics, from all areas of peptide chemistry. Worth looking at, with a number of really useful chapters.

Peptide Literature

International Journal of Peptide and Protein Research (*Int. J. Pept. Protein Res.*). This journal is specifically devoted to the chemistry of peptides and proteins, and contains many key papers. But much of peptide chemistry is also published in the general/organic chemistry journals.

Amino Acids, Peptides and Proteins, Volumes 1–16, and *Amino Acids and Peptides*, Volumes 17– , Specialist Periodical Reports published by the Royal Society of Chemistry. These invaluable reports are published annually, and summarise the previous year's literature in the field. After 1983 (Volume 16), proteins were no longer included in the Reports, because of the sheer number of papers being published.

Single-volume Texts

Comprehensive Organic Chemistry, Volume 5 (E. Haslam, Ed.), Part 23, *Proteins: Amino-acids and Peptides*, pp. 177–385, Pergamon Press, 1979. This major series of organic chemistry books contains a useful section on amino acids, peptide synthesis, and peptide structure.

MTP International Review of Science, Volume 6, *Amino Acids, Peptides, and Related Compounds*, Butterworths. This review of science series has a useful volume devoted to developments in the key aspects of peptide chemistry.

Perspectives in Peptide Chemistry (A. Eberle, R. Geiger, and T. Wieland, Eds.), Karger, 1981. A series of short chapters on almost all aspects of peptide chemistry and biochemistry. Really worth browsing through, with some excellent contributions from the experts.

CHAPTER 2

Amino Acids

Naturally occurring peptides are extremely diverse in their structure and properties. For many peptides, the sequence of the amino acids is controlled by the genetic code; and, perhaps surprisingly, there are only 20 amino acids that are DNA encoded. Nevertheless, for a simple pentapeptide, there are a total of 3,200,000 (i.e. 20^5) possible combinations. This allows nature to have an almost limitless array of peptides and proteins. Despite this, there are also a large number of peptides in nature that contain amino acids that are not DNA encoded, including several important antibiotics (e.g. penicillin precursors).

In this chapter, we will look initially at the DNA encoded amino acids, and then at some of those that are not DNA encoded. We will consider next the chemical reactions that amino acids undergo, and finally we will discuss their physical properties.

1 The DNA Encoded Amino Acids

1.1 General Structure

All of the 20 DNA encoded amino acids are α-amino acids, which means that the amino and carboxylic acid groups are both attached to the same carbon atom. Therefore they all possess the same generalised structure shown below, and the only difference between them is the nature of the R-group.

$$\underset{H_2N}{\overset{\displaystyle R}{\underset{\displaystyle}{\overset{|}{\underset{\displaystyle CH}{\diagup\diagdown}}}}}CO_2H$$

The R-group on an amino acid or on a residue in a peptide is known as the side-chain.	The α-carbon of an amino acid is the one to which the carboxylic acid group is attached. For α-amino acids, the amino group is attached to this α-carbon.

Another feature of DNA encoded peptides is that the constituent amino acids are **always** covalently bonded via the α-amino and α-carboxylic acid groups; in other words, irrespective of the side-chain, the peptide bonds are only formed between the α-groups of the constituent amino acids.

1.2 Nature of the Side-chain

The 20 possible side-chains for the DNA encoded amino acids are summarised in the following table. The name and three-letter abbreviation for each of the amino acids are also given; there is no need to learn them, although there is another table of them at the back of this book, for rapid reference. Some of the amino acids will crop up quite often, and so you may find that you begin to remember their structures anyway.

Table 2.1. The 20 DNA encoded amino acids.

Name	Side-chain	Abbrev.	Name	Side-chain	Abbrev.
Alanine	CH_3-	Ala	Leucine	$(CH_3)_2CH-CH_2-$	Leu
Arginine	$\underset{H_2N}{\overset{HN}{>}}C-NH(CH_2)_3-$	Arg	Lysine	$H_2N-(CH_2)_4-$	Lys
Asparagine[†]	$H_2NCO-CH_2-$	Asn	Methionine	$CH_3S-(CH_2)_2-$	Met
Aspartic acid	HO_2C-CH_2-	Asp	Phenylalanine	$\langle O \rangle-CH_2-$	Phe
Cysteine	$HS-CH_2-$	Cys	Proline*	(ring)$-CO_2H$	Pro
Glutamic acid	$HO_2C-(CH_2)_2-$	Glu	Serine	$HO-CH_2-$	Ser
Glutamine[†]	$H_2NCO-(CH_2)_2-$	Gln	Threonine	$CH_3CH(OH)-$	Thr
Glycine	$H-$	Gly	Tryptophan	(indole)$-CH_2-$	Trp
Histidine	(imidazole)$-CH_2-$	His	Tyrosine	$-\langle O \rangle-CH_2-$	Tyr
Isoleucine	$C_2H_5-CH(CH_3)-$	Ile	Valine	$(CH_3)_2CH-$	Val

*Proline is drawn in full, to show how the side-chain is actually joined to the α-amino group. This is the only secondary amino acid to occur naturally in proteins.
[†]Asparagine and glutamine are simple derivatives of aspartic acid and glutamic acid respectively, in which the side-chain is present as the amide (neutral) rather than as the free acid.

Although we are looking at the properties of **amino acids** in this chapter, it is useful to remember that any properties of the **side-chain** will still be valid, even if the amino acid is part of a **peptide**. In contrast, the α-amino and α-carboxylic acid groups that form the peptide bonds will be present as the relatively unreactive amide moiety. So the properties of the side-chains of the constituent amino acids dominate the properties of peptidic molecules.

The carboxylic acid group of an amino acid no longer possesses an acidic hydrogen when it is part of a peptide bond.	The amino group of an amino acid is no longer basic when part of a peptide bond because the lone pair is involved in conjugation with the carbonyl group.

For convenience, the amino acids in the table have been placed in alphabetical order. This is not particularly logical, because the side-chains modify the properties of the molecules considerably. For example, the side-chain of lysine contains an extra functional group.

Lysine

As you can see, there is an additional amino group present. Thus, even when the α-amino group is involved in a peptide bond, lysine will remain basic because of the additional NH_2-group on the side-chain; these types of molecules are called **basic amino acids**.

Similarly, glutamic acid has a carboxylic acid group on its side-chain. When the glutamic acid residue is present in a peptide, the α-carboxylic acid group will no longer be acidic if it is involved in a peptide bond. But an acid function will still be present on the side-chain, and these types of molecules are called **acidic amino acids**. (Remember, only the α-amino and α-carboxylic acid groups are involved in the peptide bond—the side-chain is not generally modified.)

Glutamic acid   **Cysteine**

The amino acid cysteine has a thiol group in its side-chain. When thiols are gently oxidised (with air, or iodine), an S—S bond is formed.

$$\text{i.e.} \quad R—SH \xrightarrow{\text{Oxidise}} RS—SR$$

Cysteine is no exception to this chemical reaction, and cysteine residues readily form sulphur–sulphur bonds between each other; this S—S linked dimer is often called cystine. This is a very important feature of many peptides and proteins— these S—S bonds are often formed between cysteine residues that are far apart in the primary sequence, but that can be close together in space; this limits the possible conformations for peptides, and is crucial to the well-defined three-dimensional structure of many proteins.

So, in summary, all amino acids possess an amino group, a carboxylic acid group, and a side-chain that can significantly alter their properties. However, the DNA encoded amino acids also share one other very important feature.

1.3 Stereochemistry

Despite the wide variety of possible side-chains, all of the common amino acids have the same stereochemistry at the α-carbon (marked ∗ below).

Stereochemistry of the DNA encoded amino acids

These DNA encoded amino acids are usually called L-amino acids. This is rather an old-fashioned nomenclature, and refers to the correlation between the compound and L-glyceraldehyde.

L-**Amino acid**   L-**Glyceraldehyde**

The modern nomenclature usually employs the prefixes R and S.

***

The R/S notation. Each of the four groups surrounding a chiral atom are given an order of precedence based on their atomic number, or on that of the next atom(s) along. Thus, for example, NH_2 (AN of $N = 7$) is before CO_2H (AN of $C = 6$); and CO_2H (carbon bonded to three oxygens, as the double bond counts twice) would come before CH_2OH (carbon bonded to one oxygen); these in turn would come before any alkyl group, which would take precedence over hydrogen. The molecule is then viewed from the far side of the lowest-priority group (usually hydrogen), and the three remaining groups are considered in order of precedence; if their sequence (i.e. highest to lowest priority) is clockwise then the molecule is termed R, whereas if it is anti-clockwise the molecule is of the S-configuration.

***

If we apply the R/S rules to the DNA encoded amino acids, we can work out whether they should be classed as being of the S- or of the R-configuration. The Newman projection below shows the molecule being viewed from the far side of the hydrogen atom.

L-**Amino acid**   **Newman projection**

Most of the common amino acids are defined as being S, with the order of priority being $NH_2 > CO_2H > R > H$. There are, however, a couple of exceptions:

(i) Glycine. The side-chain of this amino acid is simply a hydrogen atom. This means that glycine is not chiral, and it has no optical isomers.

(ii) Cysteine. Because the side-chain is CH_2SH (carbon bonded to sulphur), this group takes precedence over CO_2H (carbon bonded to oxygen), and this molecule is designated as being R.

The mirror image of an optical isomer is called its **enantiomer**. These enantiomers react identically in most chemical environments.

L-**Amino acid**   **Mirror**   D-**Amino acid**

However, these enantiomers do rotate polarised light in opposite directions, and can also be distinguished by other optically active molecules. In particular, enzymes are composed of only L-amino acids, and can usually **only** interact with molecules of the correct absolute stereochemistry.

It is very important to realise that **all** of the DNA encoded amino acids have the same layout in three dimensions. There are no exceptions, although nature (and man) has been able to produce a number of important D-amino acids with reversed stereochemistry.

1.4   Amino Acid and Peptide Shorthand

As you will now know, all of the common amino acids have simple three-letter abbreviations (see the table on page 12). These shorthand versions are very convenient, and the full names of the corresponding amino acids are usually quite obvious (e.g. **Gly** is **gly**cine, **Phe** is **phe**nylalanine, etc.). But the real advantage of the three-letter abbreviations is for concisely expressing the structure of peptides or substituted amino acids.

Amino acids are each given a three-letter symbol. A dash **before** the symbol indicates a bond to the α-**nitrogen** of the amino acid. A dash **after** the symbol indicates a bond to the α-**carbonyl** group of the carboxylic acid.

e.g.

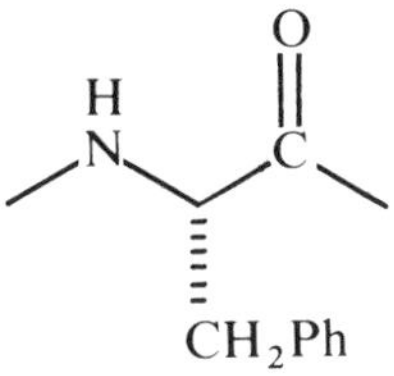

$\equiv$ **—Phe—**

So, if the α-amino group of lysine were acetylated (ethanoylated), then this could be written as:

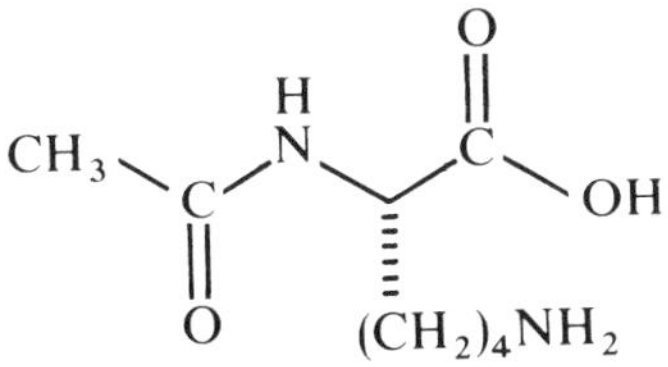

$\equiv$ **Ac—Lys** (NB: Ac $\equiv$ CH_3CO)

Similarly, if the carboxylic acid group of lysine were transformed into the methyl ester, then the product could be abbreviated to:

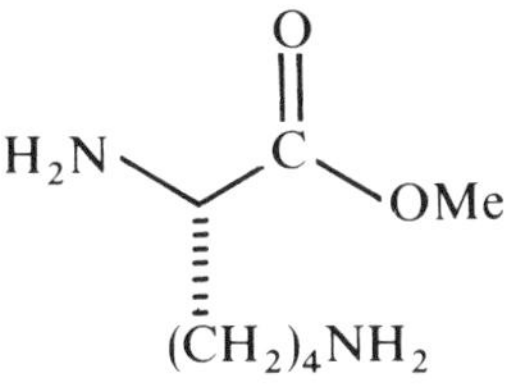

$\equiv$ **Lys—OMe**

If an amino acid has a modification to the side-chain, then this is shown in brackets, or through a vertical dash. For example, if the side-chain amino group of lysine were acetylated, then this could be written as:

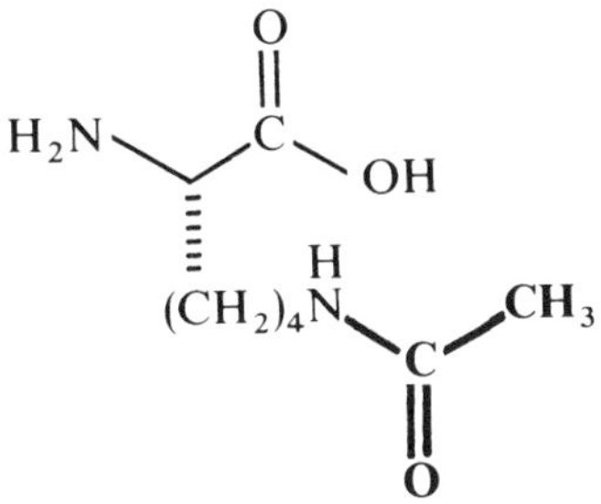

$\equiv$ **Lys(Ac)** or **Lys**

If **two** amino acids are joined by a peptide bond, then this is designated by a simple dash; so a dipeptide of glycine and phenylalanine could be written as:

Gly—Phe or Phe—Gly

However, these two peptides are *not* the same. Peptides (like their constituent amino acids) are always written with the N-terminal at the left, and the C-terminal at the right:

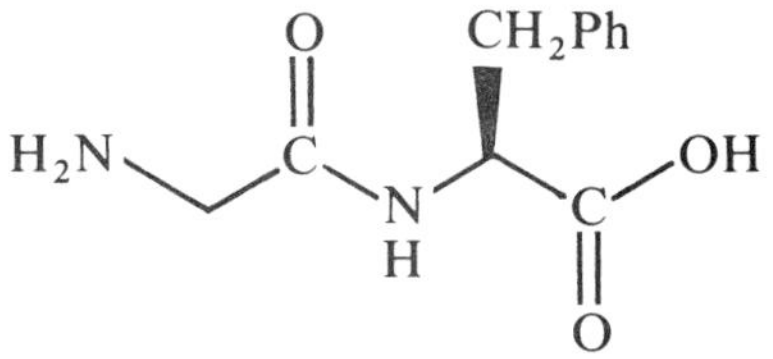

For some cyclic peptides, it can be unclear which way the peptide bonds are going. In these cases, an arrow on the dash indicates the direction of the peptide bond (N → CO).

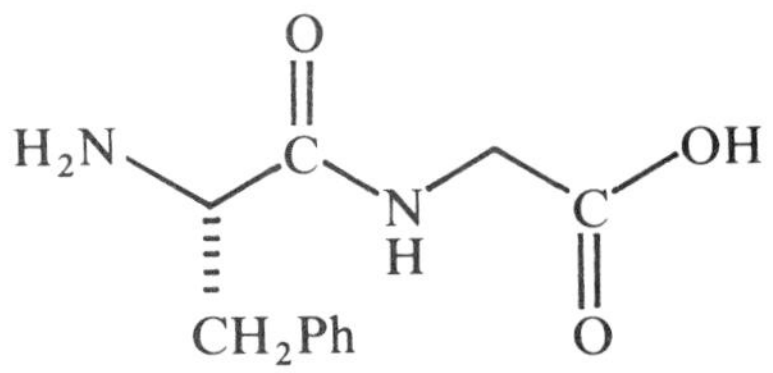

If the sequence of the residues in a peptide is not known, then the amino acid symbols are separated by commas. For example, [**Gly(2), Lys, Phe**] represents a tetrapeptide that is composed of **two glycines, one lysine,** and **one phenylalanine**— but no particular order for the residues in the peptide is implied.

Finally, unless otherwise stated, it is always assumed that abbreviations for amino acids or residues correspond to the L-configuration.

2 Unusual Amino Acids

There are an enormous number of naturally occurring amino acids that are not DNA encoded—here they are termed 'unusual' amino acids simply because they are rather less abundant that those encoded by DNA. Furthermore, there are many amino acids that have been synthesised for incorporation into peptides in order to produce modified biological properties. The diversity of naturally occurring amino acids is exemplified below, with the molecules being grouped into structural types.

(a) Unusual side-chains

Examples of naturally occurring amino acids with unusual side-chains

Other examples include ornithine and hydroxyproline which are *not* DNA encoded, although the structurally related lysine and proline *are* DNA encoded. Ornithine is present in many naturally occurring peptides, including several antibiotics. Hydroxyproline is a major constituent of collagen, and is formed by the oxidation of proline.

Ornithine   **Lysine**

Hydroxyproline   **Proline**

(b) β- and γ-Amino acids

β-Amino acids	γ-Amino acids

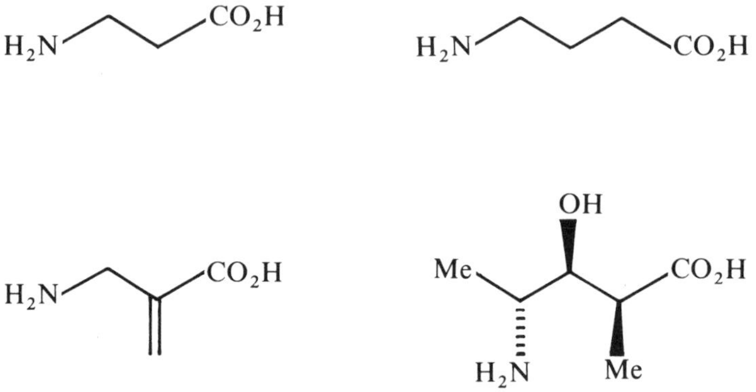

Examples of naturally occurring β- and γ-amino acids

(c) D-Amino acids

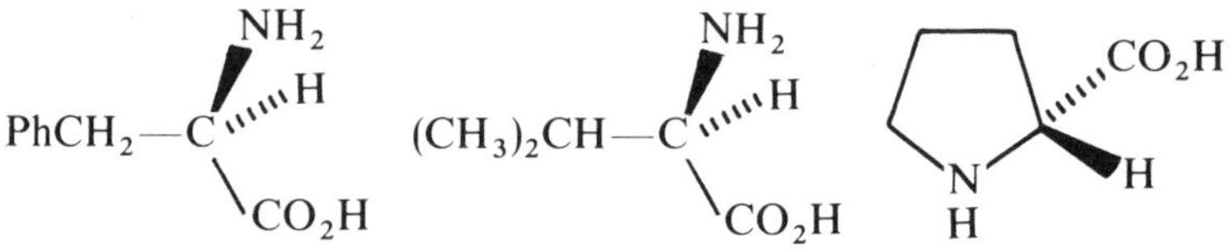

D-Phe   **D-Val**   **D-Pro**

Examples of naturally occurring D-amino acids

Phenylalanine, valine and proline are all of the L-configuration when DNA encoded. However, the D-isomers are synthesised by several organisms, and they are vital constituents of a number of antibiotics.

(d) Other naturally occurring non-DNA encoded amino acids

$$Me_3\overset{\oplus}{N}CH_2CO_2^{\ominus} \qquad cf. \qquad H_2NCH_2CO_2H$$

Betaine   **Glycine**

Betaine is a methylating agent, used in the biosynthesis of methionine. D-α-Aminoadipic acid is one of the constituent amino acids of the Cephalosporin antibiotics (and of some Penicillins) and has both an unusual side-chain (cf. glutamic acid), and is of the D-configuration.

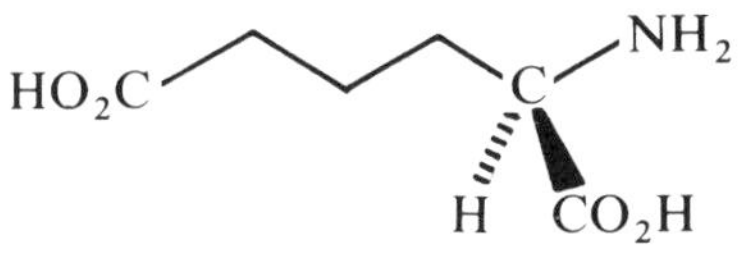

D-α-Aminoadipic acid   cf.   **Glutamic acid**

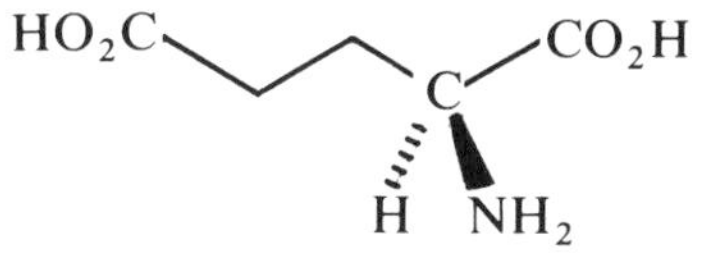

In order to illustrate the importance of some of these compounds, here are three examples of biologically active molecules that are composed of "unusual" amino acids.

(i)

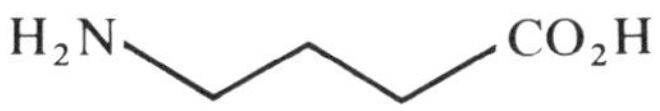

GABA

γ-Aminobutyric acid (GABA) is involved in the transmission of nerve impulses. It is formed by the decarboxylation of glutamic acid, and reduces the sensitivity of nerve cells to stimulation.

(ii)

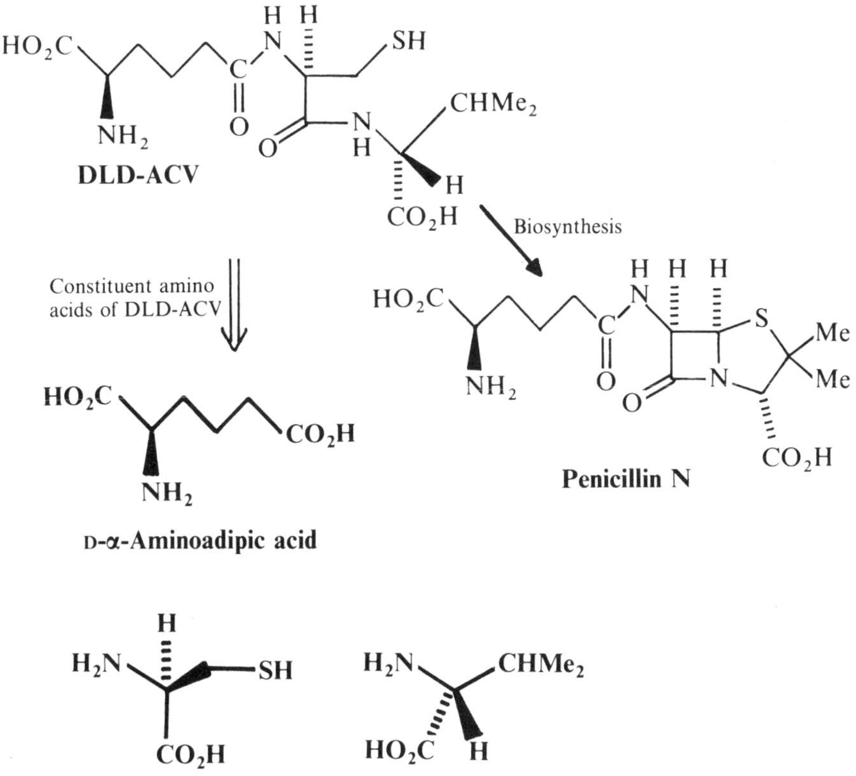

DLD-ACV

Constituent amino acids of DLD-ACV

Biosynthesis

Penicillin N

D-α-Aminoadipic acid

L-Cysteine   **D-Valine**

22

Penicillins are biosynthesised from tripeptides such as DLD-ACV, as shown on page 21. The constituent amino acids of DLD-ACV are also shown, but the mechanism by which Penicillins are biosynthesised from the tripeptide precursor is still not fully understood.

(iii)

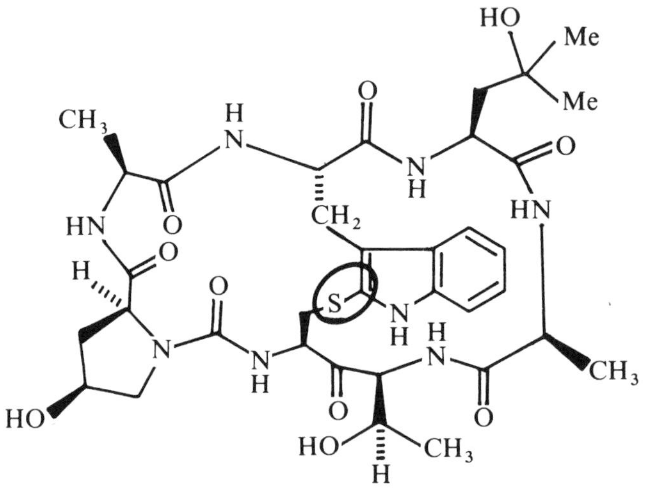

Phalloin

Phalloin is composed almost entirely of normal DNA encoded L-amino acids (the proline has an extra *trans* OH group, and one of the threonines has an extra Me group). What is unusual about this peptide is the intramolecular linkage between Cys and Trp; this creates a bicyclic heptapeptide, which is one of several extremely toxic compounds made by the Deathcap toadstools.

3 Chemical Reactions of Amino Acids

We will consider separately the chemical properties of the α-amino and the α-carboxylic acid groups, before looking at reactions that involve both of them. Then we will discuss the chemical reactions that the amino acid side-chains might undergo.

3.1 The α-Amino Group

There is nothing unusual about the properties of the α-amino group. Like most amines, it can act either as a base or as a nucleophile. For example, it can be readily protonated or methylated:

e.g. $R{-}NH_2 \xrightarrow{HCl} R{-}NH_3^{\oplus} + Cl^{\ominus}$

e.g. $R{-}NH_2 \xrightarrow{MeI} R{-}NHMe \xrightarrow{Excess\ MeI} R{-}NMe_3^{\oplus}I^{\ominus}$

More importantly, amines can readily attack even moderately reactive carbonyl groups. For example, aldehydes and ketones yield imines (although these can usually be hydrolysed readily back to the starting materials), whilst acid chlorides or acid anhydrides generate amide derivatives.

e.g. $R{-}NH_2 \xrightarrow{R^1COR^2} R{-}N{=}CR^1R^2$

e.g. $R{-}NH_2 \xrightarrow{R^1COCl} R{-}NHCOR^1$

e.g. $R{-}NH_2 \xrightarrow{(R^1CO)_2O} R{-}NHCOR^1$

These last two types of reaction are vital to the synthesis of peptides, and we will come back to this kind of chemistry in Chapter 6.

Amino groups can also react with electrophilic sulphur reagents. For example, in the reaction with 4-methylbenzenesulphonyl chloride, the corresponding sulphonamide is formed.

So, the α-amino group can react with a whole range of electrophilic reagents, giving the products that you would expect for any simple amine.

3.2 The α-Carboxylic Acid Group

Again, amino acids possess a typical carboxylic acid group. However, carboxylic acids themselves are usually rather unreactive. This is because the addition of nucleophile usually results in the removal of the labile proton, giving the carboxylate ion. (Remember, most nucleophiles are also basic to some extent, although some powerful nucleophiles are weak bases, and vice versa.)

i.e. $R{-}CO_2H \xrightarrow[(-H^{\oplus})]{Base} \left[R{-}C{\overset{\nearrow O}{\underset{\searrow O_{\ominus}}{}}} \leftrightarrow R{-}C{\overset{\nearrow O^{\ominus}}{\underset{\searrow O}{}}} \right]$

Not only is the carboxylate ion electron rich (and therefore not susceptible to nucleophilic attack), but also the charge is so spread out that it is only a very poor nucleophile itself. This means that carboxylic acids are rather unreactive. Under conditions of acid catalysis, ester formation is relatively straightforward.

Although simple esters are most readily formed under conditions of acid catalysis, more complex derivatives are better made by the addition of powerful dehydrating agents such as dicyclohexylcarbodiimide (DCCD). The use of DCCD is particularly important in the synthesis of peptides, and will be discussed in more detail in Chapter 6.

The structure of DCCD is as follows.

DCCD   **DCU**
(R = cyclohexyl = $\langle$ ◯ $\rangle$)

DCCD readily abstracts water to form dicyclohexylurea (DCU).

3.3 Reactions Involving both the Amino and Acid Functions

There are a small number of important reactions that fall into this category.

3.3.1 *Reaction with Ninhydrin.* The reaction of α-amino acids with ninhydrin is often used for their detection and quantification.

Ninhydrin acts as if it were the triketone (1) shown below, with the middle ketone group being the most reactive.

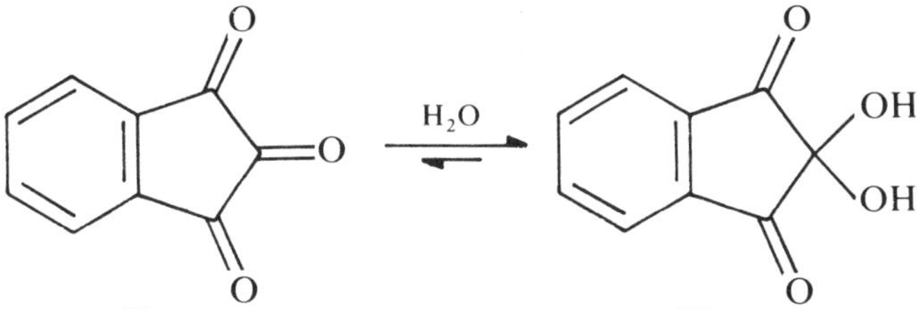

(1)   **(2)**

In fact, this middle ketone is so electron deficient that ninhydrin is stored and used as the hydrate (2). This does not affect its chemistry.

The reaction of ninhydrin with α-amino acids involves the initial formation of an imine (3), which decarboxylates to form a new imine (4); hydrolysis of (4) generates another free amino group (5), which can react with a second molecule of ninhydrin.

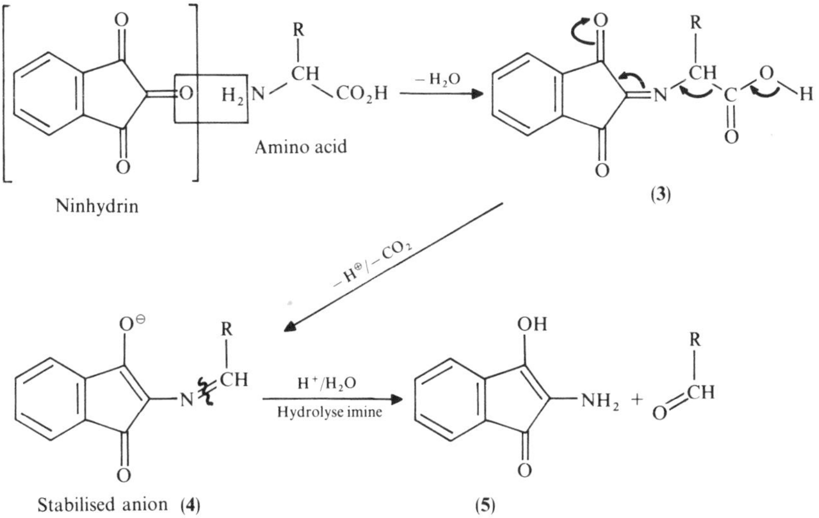

Ninhydrin   Amino acid   (3)

Stabilised anion **(4)**   **(5)**

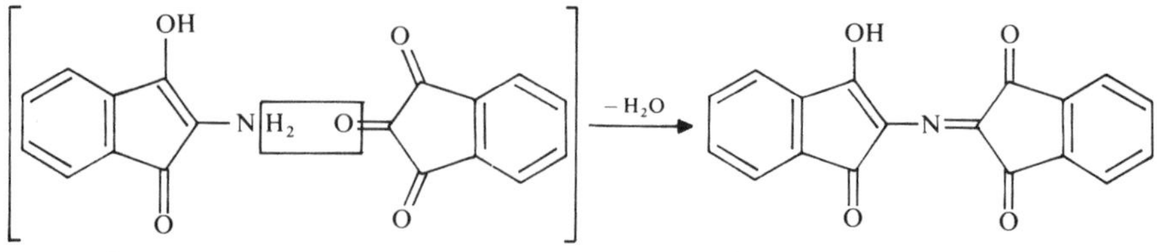

(5)   Ninhydrin λ_{max} 570 nm

The highly conjugated product absorbs light in the visible region, and is an intense purple colour (λ_{max} 570 nm). This colour is characteristic of α-amino acids, and is used for their detection. The only common α-amino acid that fails to give a positive result is proline. This is because the α-amino group is secondary, and the later stages of the reaction with ninhydrin require a primary amino function; however, a yellow colour is observed instead, which can be used for the detection of proline.

3.3.2 *De-amination Reactions.* The reaction of α-amino acids with ninhydrin is, in fact, a de-amination process; it has been given a whole section of its own because it is such an important reaction for the detection and quantification of amino acids. Aliphatic primary amines can also be de-aminated by treatment with nitrous acid (nitric III acid), with the corresponding alcohol being generated:

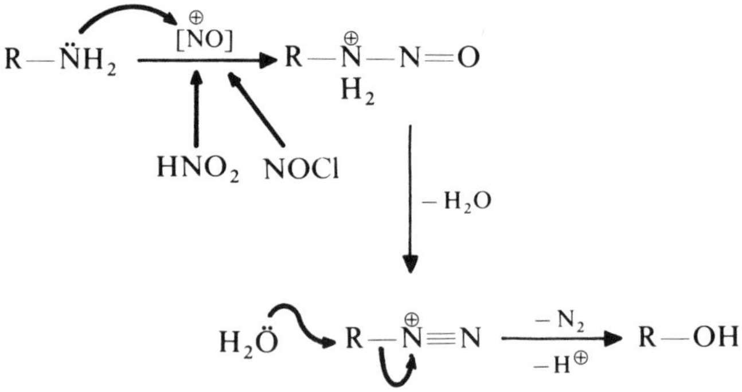

For α-amino acids, the stereochemistry of the alcohol corresponds to retention of configuration at the α-carbon, and this is due to the carboxylic acid being involved in the reaction:

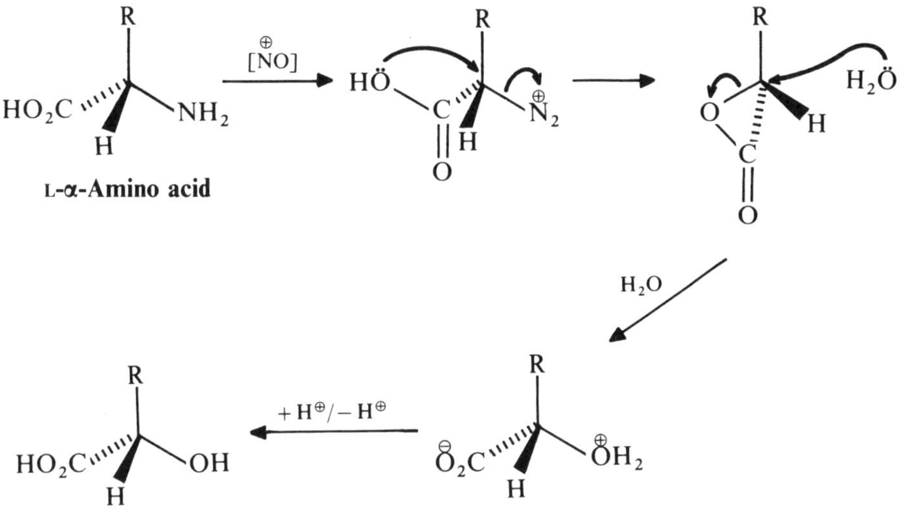

L-α-Amino acid

L-α-Hydroxy acid
(Overall retention of configuration)

De-aminations can also be initiated by the addition of pyridoxal.

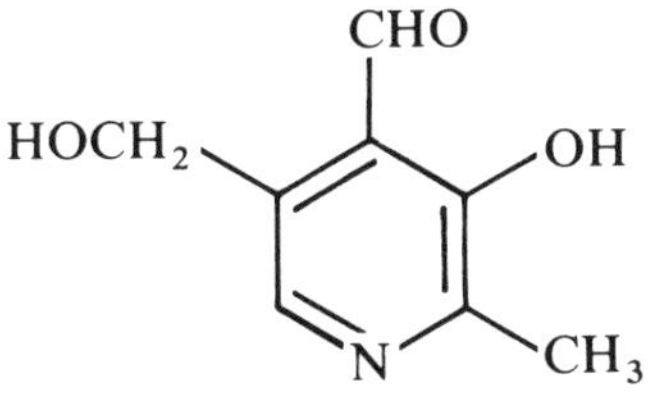

Pyridoxal

We will not go into the details of the reaction with pyridoxal, but it catalyses the conversion of $R—CH(NH_2)CO_2H$ into $R—NH_2$ or $R—CHO$ (via a mechanism that closely mirrors that with ninhydrin). Pyridoxal is present as a co-enzyme in certain biological systems, and is used for the conversion of amino acids into other biologically important molecules.

3.3.3 *Chelation Reactions.* When α-amino acids are reacted with certain metal ions, chelation complexes can be formed. For example, with Cu^{2+} ions, the following complex is formed:

Cu^{2+} chelate with amino acids

This complex is a distinctive blue colour for the glycine derivative. More importantly, complexes of this type de-activate the α-amino and α-carboxylic acid groups; this can allow the side-chain to be reacted in a selective manner, and is used for the preparation of certain protected amino acids.

3.4 Reactions of the Side-chain

These are, for the most part, the reactions that you would expect. Thus, the side-chain of lysine undergoes the typical reactions of an amine, whilst the side-chain of glutamic acid reacts as a normal carboxylic acid.

Specific tests have been developed for some residues, which exploit the chemical reactions of the side-chain. For example, the affinity of sulphur for mercury can be utilised in the selective reaction of cysteine (as the free amino acid, or as a residue in a peptide) with 4-chloromercuribenzoate:

The reagent is usually administered as the sodium salt ($M = Na$), but it is easily protonated under acidic conditions ($M = H$). The product can be readily isolated by chromatography (see later), and has a characteristically strong UV absorbance. Selective tests for several other amino acid side-chains are also known.

Another important reaction of cysteine is the oxidative dimerisation of the thiol group to give cystine; this disulphide bridge occurs in many peptides and proteins.

Cysteine   Cystine

For a few amino acids, the side-chain can actually react with the α-amino or α-carboxylic acid groups. For example, if glutamic acid is strongly heated, loss of water takes place, and a five membered cyclic amide (γ-lactam) is formed:

Glutamic acid (Glu)   Pyroglutamic acid (Glp)

There is nothing special about this chemistry—most other 4-aminobutanoic acids undergo this reaction too.

One other important property of some side-chains is that they can direct the cleavage of a specific peptide bond; this feature will be discussed in connection with the sequencing of peptides in Chapter 5.

4 Physical Properties of Amino Acids

The physical properties of amino acids will directly influence much of the chemistry that will be discussed later in this book. For example, the isolation, analysis, and conformation of peptides are all crucially dependent on the physical properties of the constituent amino acids.

In this section we will concentrate on the **factors** that influence the physical properties of amino acids—we will see in later chapters how these factors can be utilised in the study of peptides.

4.1 Ionisation of Amino Acids

The physical properties of the amino acids are dramatically influenced by the degree of ionisation at different pHs. We will look at the amino and carboxylic acid groups separately, before considering the overall ionisation, and the influence of side-chains.

4.4.1 *The α-Carboxylic Acid Group.* At pH 7, aliphatic carboxylic acids are normally present as the carboxylate ion. In other words, the following equilibrium would lie to the right.

$$RCO_2H + H_2O \rightleftharpoons RCO_2^- + H_3O^+$$

Generally
$$AH + H_2O \rightleftharpoons A^- + H_3O^+$$

$$K = \frac{[RCO_2^-][H_3O^+]}{[RCO_2H][H_2O]}$$

Equilibria in aqueous solution. In aqueous solution, H^+ is hydrated, and is often written as H_3O^+; either abbreviation (H^+ or H_3O^+) for the hydrated proton is acceptable. Moreover, in aqueous equilibria, the $[H_2O]$ (which is in such huge excess that it remains effectively constant at about 55M) is usually incorporated into the equilibrium constant, K. In practice, this means that the $[H_2O]$ can be ignored, but the equilibrium constant always has a subscript (K_a or K_b) or prime (K'), to indicate that $[H_2O]$ is already taken care of.

So, for the dissociation of our carboxylic acid (where $[H_2O]$ is effectively constant) K_a is usually quoted (where $K_a = K \times [H_2O]$).

$$\text{i.e. } K_a = \frac{[RCO_2^-][H_3O^+]}{[RCO_2H]}$$

The actual value for the equilibrium constant for this reaction would typically be about 10^{-3}, so the pK_a would be about 3.

For our carboxylic acid in aqueous solution, if we adjusted the pH to 3, then:

$$K_a = [A^-] \times 10^{-3}/[AH] = 10^{-3}$$

So
$$[A^-] = [AH]$$

Equilibrium constants (K) involving acids and bases are often expressed as the pK or $-\log_{10}K$. Using this nomenclature, pH is a measure of $-\log_{10}[H^+]$.

For aqueous solutions, the equilibrium constant for

$$H_2O \rightleftharpoons H^+ + HO^-$$

is 10^{-14}, i.e.

$$K_a = [H^+] \times [HO^-] = 10^{-14}$$

For neutral solutions, the concentrations of $[H^+]$ and $[HO^-]$ must be equal; so, $[H^+] = [HO^-] = 10^{-7}$. Thus, the pH of neutral water is 7.

So at pH 3, we can see that our carboxylic acid would be exactly **half** ionised.

At pH 7, the $[H^+]$ would be 10^4 times lower. This would pull the equilibrium over to the right, so that $10^4 (= 10,000)$ times as much of the carboxylate would be deprotonated than would be protonated.

i.e.

$$[A^-] \times 10^{-7}/[AH] = 10^{-3}$$

So

$$[A^-]/[AH] = 10^4$$

In other words, 99.99% of the acid would be present as the carboxylate, and only 0.01% would be present in the protonated form, at pH 7.

4.1.2 *The Amino Group.* At pH 7, most aliphatic amines are protonated. In other words, the following equilibrium lies to the right:

$$R-NH_2 + H_2O \rightleftharpoons R-NH_3^+ + HO^-$$

Generally

$$B + H_2O \rightleftharpoons BH^+ + HO^-$$

$$K = \frac{[R-NH_3^+][HO^-]}{[R-NH_2][H_2O]}$$

This equation leads to a modified equilibrium constant K_b because it refers to basicity (rather than K_a for acidity):

i.e.

$$K_b = K \times [H_2O] = \frac{[R-NH_3^+][HO^-]}{[R-NH_2]}$$

A typical value of K_b for an amine is 10^{-5} (i.e. p$K_b = 5$). So if the pH were altered so that $[HO^-]$ were 10^{-5}, then;

$$K_b = [BH^+] \times 10^{-5}/[B] = 10^{-5}$$

So

$$[BH^+] = [B]$$

It would be very helpful to know the actual value of the pH at which the amine is half ionised; i.e. the pH at which $[HO^-] = 10^{-5}$.

We can convert $[HO^-]$ into pH very easily. Under aqueous conditions, we know that:

$$[H^+][HO^-] = 10^{-14}$$

It therefore follows that:

$$pH + pOH = 14$$

(where p means $-\log_{10}$)
So when $[HO^-] = 10^{-5}$, then

$$pOH = -\log_{10}[HO^-] = 5.$$

Hence pH $= 14 - 5 = 9$

So, at pH 9, we can see that our amine would be exactly half ionised.

---

pK_a *and* pK_b. It is often considered to be more convenient to quote the pK_a value for an amine (i.e. the pH at which it is half ionised) rather than the pK_b value. But in fact, it can easily be shown that

$$pK_a + pK_b = 14$$

so pK_a and pK_b can be readily interconverted anyway.
We will use $pK_a(\geqslant NH^+)$ to indicate the pH at which amine/ammonium ion are at equal concentrations.

So the pK_b for a typical amine is about 5, indicating a $pK_a(\geqslant NH^+)$ of 9, and hence half-ionisation at pH 9. Hence, at pH 7, $[H_3O^+]$ would be 10^2 ($= 100$) times higher, so there would be 100 times as much protonated amine as free amine.

As

$$[BH^+] \times [HO^-]/[B] = K_b$$

Then

$$[BH^+] \times 10^{-7}/[B] = 10^{-5} \text{ (at pH 7)}$$

Hence

$$[BH^+]/[B] = 10^2$$

In other words, 99% of the amine would be protonated, and only 1% would be present as the free amine, at pH 7.

4.1.3 *Amino Acids.* With an amino and a carboxylic acid group being present in these molecules, there is both a basic and an acidic component in them. Both functional groups could be ionised as discussed above.

So at pH 7, the carboxylic acid group would be almost totally deprotonated (as the carboxylate) whilst the amino group would be more than 99% protonated (as the ammonium ion). Indeed, anywhere between pH 4 and pH 8, over 90% of the amino acid molecules would have both of the α-functional groups ionised. (In fact, the α-amino group slightly increases the acidity of the carboxylic acid group, and the carboxylic acid group slightly increases the basicity of the α-amino group, due to inductive effects.)

$$
\begin{array}{c}
R \\
| \\
CH \\
{}^{\ominus}O_2C \diagup \quad \diagdown \overset{\oplus}{N}H_3
\end{array}
$$

Zwitterion

> Molecules that possess both positively and negatively charged groups are called **zwitterions**.

We can summarise the ionisation characteristics of amino acids by sketching the titration curves of (for example) glycine against sodium hydroxide solution (Figure 2.1, page 34).

At all the pHs shown in the titration curve, glycine has at least one functional group that is almost totally ionised—and this is true of most amino acids. The presence of these charged groups does not greatly affect the chemical reactions of amino acids (because the ionisation steps are equilibria); but it does have a big influence on their physical properties. For example, unlike most organic molecules of moderate molecular weight, the amino acids are involatile solids, which have high melting points (generally > 200 °C); this is because the zwitterions can form strong ionic interactions between each other. Indeed, most of the physical properties of amino acids (and peptides) that are used for their purification take advantage of the degree of ionisation of the molecules at different pHs. We will therefore have to look at the ionisation of amino acids in a little more detail, and in particular we will need to consider the effect of the side-chains.

4.1.4 *Influence of the Side-chain.* There are two possible effects that the side-chain can have, as far as the ionisation of amino acids is concerned.

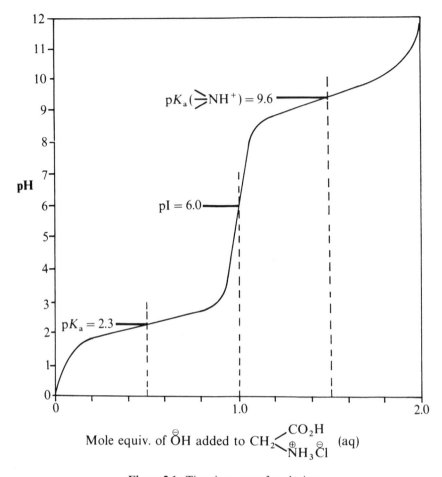

Figure 2.1. Titration curve for glycine.

(a) The side-chain could have a simple inductive effect, which would slightly alter the acidity and basicity of the α-groups.

For example, the phenyl group of phenylalanine causes the side-chain to be slightly electron withdrawing. This makes the carboxylic acid **more** acidic ($-I$ effect stabilises the carboxylate ion), but the amino group **less** basic ($-I$ effect destabilises the ammonium ion). Therefore both the pK_a and the pK_b for phenylalanine will be lower than for glycine. This can be seen quite clearly if we compare the titration curves for phenylalanine and glycine (Figure 2.2).

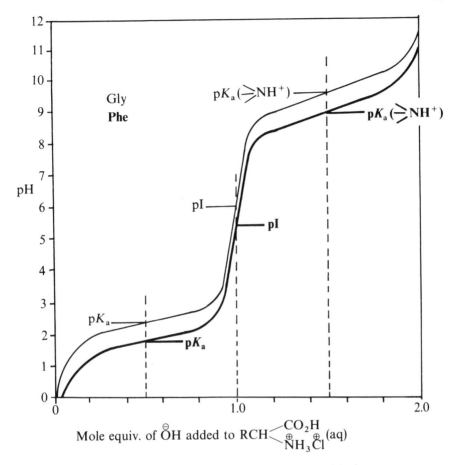

Figure 2.2. Titration curves for glycine and phenylalanine.

(b) The side-chain could itself possess an acidic or basic group. This would have a dramatic effect on the overall ionisation of the molecule at different pH values.

For example, because of the carboxylic acid group on the side-chain of glutamic acid, this amino acid would possess **two** carboxylates and **one** ammonium ion at pH 7—and would therefore carry an overall negative charge. Conversely, the amino acid lysine possesses an additional amino group on the side-chain, and would carry an overall positive charge at pH 7.

The overall charge on an amino acid, at given pHs, is one of the main properties used in the separation of mixtures of amino acids (see Chapter 4).

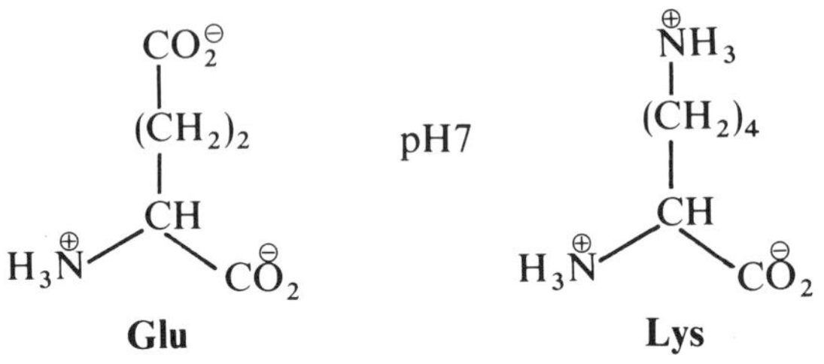

$$
\begin{array}{cc}
\underset{\substack{\displaystyle CO_2^{\ominus} \\ | \\ (CH_2)_2 \\ | \\ CH \\ H_3\overset{\oplus}{N}\diagup \quad \diagdown CO_2^{\ominus}}}{\mathbf{Glu}}
& pH7 &
\underset{\substack{\displaystyle \overset{\oplus}{N}H_3 \\ | \\ (CH_2)_4 \\ | \\ CH \\ H_3\overset{\oplus}{N}\diagup \quad \diagdown CO_2^{\ominus}}}{\mathbf{Lys}}
\end{array}
$$

In order to have no overall charge on glutamic acid, we would need to lower the pH, until half of the carboxylate groups were protonated. Similarly for lysine, only by raising the pH would we be able to deprotonate half of the ammonium groups, and thereby generate a molecule with no net charge.

The pH at which an amino acid (or peptide) carries no net charge is called the **isoelectric point (pI)**. This property is widely used in purification processes.

Another way to see what is going on is to look at the titration curves; we can see that glutamic acid has an additional end-point (compared with glycine) which corresponds to the ionisation of the carboxylic acid group of the side-chain (Figure 2.3). Similarly, lysine has an additional end-point that corresponds to the deionisation of its ammonium side-chain (Figure 2.4); the isoelectric points are also shown.

Finally we can compare the net charge on each of the four amino acids whose titration curves are shown above (Figures 2.1–2.4), at different pHs. This type of analysis will give important insight into how the following amino acids could be readily separated from each other:

Glycine	(Gly)
Lysine	(Lys)
Glutamic acid	(Glu)
Phenylalanine	(Phe)

Figure 2.5 below shows the predicted charge on each of these four amino acids between pH 0 and 12.

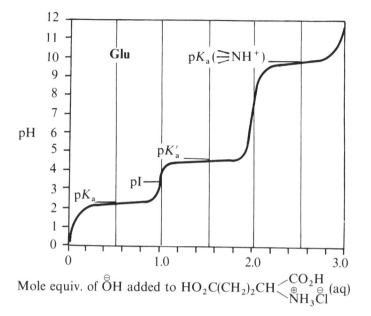

Figure 2.3. Titration curve for glutamic acid.

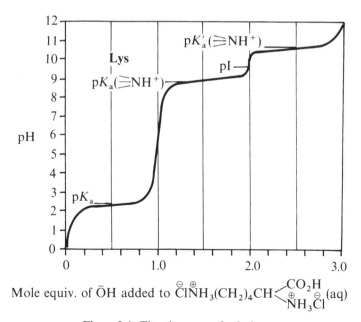

Figure 2.4. Titration curve for lysine.

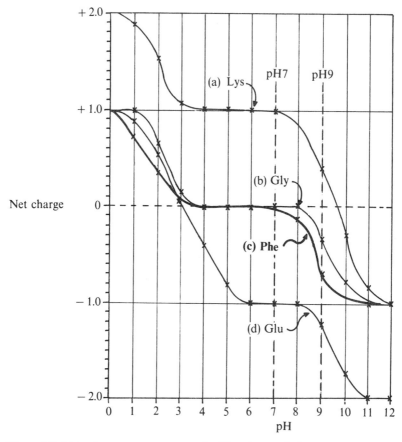

Figure 2.5. Net charge, at a range of pH values, of the amino acids (a) lysine, (b) glycine, (c) phenylalanine, and (d) glutamic acid.

Check that you agree with the approximate shape of the four curves. Two pHs have been highlighted on the graphs, in order to exemplify the differences in the net charge on the four amino acids:

> **pH 7.** Whilst Gly and Phe carry no net charge at this pH (this is close to their pIs), Glu would be negatively charged, and Lys would be positively charged.
> **pH 9.** At this pH, all four of the amino acids carry a net charge—but the charge is different for each of them. For example, the negative charge on Phe (-0.6) is about twice that on Gly (-0.3).

So it should now be clear that the amino acids carry slightly different net charges at any given pH, and this property can be used in the separation of amino acids. Furthermore, the pH at which they carry no net charge is different for each amino acid, and this completely different characteristic is also used in the purification of amino acids and peptides.

4.2 Hydrophilicity/Lipophilicity

There are other ways in which the side-chains of amino acids can influence their physical properties. Suppose that we compare the solubilities of glycine, phenylalanine, and serine in various solvents.

For a start, the fact that they all tend to exist as the zwitterions would mean that they would be much more soluble in polar solvents than in non-polar solvents. This is why amino acids (and peptides) are usually soluble in water, but are often difficult to dissolve in organic solvents. However, the **relative** solubilities of Gly, Phe and Ser in different solvents depends crucially upon the nature of the side-chains:

> (i) In **water**, which is polar and capable of hydrogen bonding, the solubility would be:
>
> Ser ($R = CH_2OH$) > Gly ($R = H$) > Phe ($R = CH_2Ph$)
>
> (ii) In **trichloromethane**, which is fairly polar, Gly, Phe and Ser would be about equally soluble.
>
> (iii) In **benzene**, which is non-polar, the solubilities would be:
> Phe > Gly > Ser

You can see that the non-polar side-chain of Phe will increase its solubility in non-polar solvents; conversely, the polar side-chain of serine will improve its solubility in polar solvents, particularly if hydrogen bonding is also possible—as in aqueous solutions. The side-chains of amino acids are said to be hydrophilic or lipophilic, and this can greatly affect their partition between two phases.

> **Hydrophilic** is from the Greek, and means 'water-loving'. These amino acids have a high affinity for water. Amino acids with ionised side-chains (e.g. Glu or Lys), or with side-chains capable of hydrogen bonding (e.g. Ser) are said to be hydrophilic.
> **Lipophilic** is also from the Greek, and means 'fat loving'. The side-chains of lipophilic amino acids contain simple alkyl or aromatic groups (e.g. Val or Phe), and have a preference for non-polar environments. The term **hydrophobic** ('water hating') is sometimes used instead of lipophilic.

The difference in the hydrophilicity/lipophilicity of the side-chains of different amino acids is rarely exploited for their separation or purification. It is, however,

40

a major characteristic used for the isolation of peptides, and is also a critical factor in the folding of peptides and proteins.

5 Summary

There are only **20 amino acids** that are **DNA encoded**, and these are given three-letter abbreviations (see the table on page 12); all of these amino acids have the same layout in three dimensions, and differences in their properties depend on the nature of the side-chain (R).

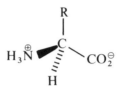

DNA encoded amino acids at pH7

There are also a whole range of **unusual amino acids** that are **not DNA encoded**. Amino acids display the expected **chemical properties** of **carboxylic acids** and of **amines**; the side-chain can also give distinctive reactions. In addition, all primary α-amino acids react with ninhydrin, giving a characteristic purple colour (λ_{max} 570 nm).

Amino acids exist as **zwitterions** at pH 7, in which both the amino and carboxylic acid groups are almost totally **ionised**; the physical properties of the amino acids are largely governed by the degree of ionisation at different pHs. The side-chain of an amino acid can alter its physical properties by modifying:

(i) The net charge at a given pH.
(ii) The pH at which there is no net charge (the isoelectric point, or pI).
(iii) The relative affinity for water.

> N.B. **Amino acids** are usually represented in their **neutral state** [i.e. $R—CH(NH_2)CO_2H$], unless the charge is being emphasised. This convention will be used throughout this book.

The chemical and physical properties of amino acids are utilised in the separation of mixtures of them, for characterising them, and in developing methods for chemically bonding them together in peptide synthesis. Of course, **peptides** are themselves composed of covalently bonded amino acids; their properties will therefore be dominated by the nature of the side-chains on the constituent amino acids.

So, having looked at some of the properties of amino acids, we can now concentrate on our (imaginary) peptide PENTIN, which we believe might be used to prevent the progression of arthritis in joints. Our first task is to obtain a sample of pure PENTIN and the purification/isolation of peptides is the topic covered in Chapter 3.

Further Reading

Chemistry and Biochemistry of the Amino Acids (G.C. Barrett, Ed.), Chapman and Hall, 1985. An extremely thorough (and readable) coverage of the structure, synthesis, and properties of amino acids.
Comprehensive Organic Chemistry, Volume 2 (I.O. Sutherland, Ed.), Part 9.6, pp. 815–40, Pergamon Press, 1979. A very useful section in this major series of textbooks.

Questions

1. (a) Define each of the following amino acids as *R* or *S*.
    (b) Which of them are L-amino acids?
    (c) Which of them are DNA encoded?

2. Chlamydocin is a naturally occurring cyclic tetrapeptide:

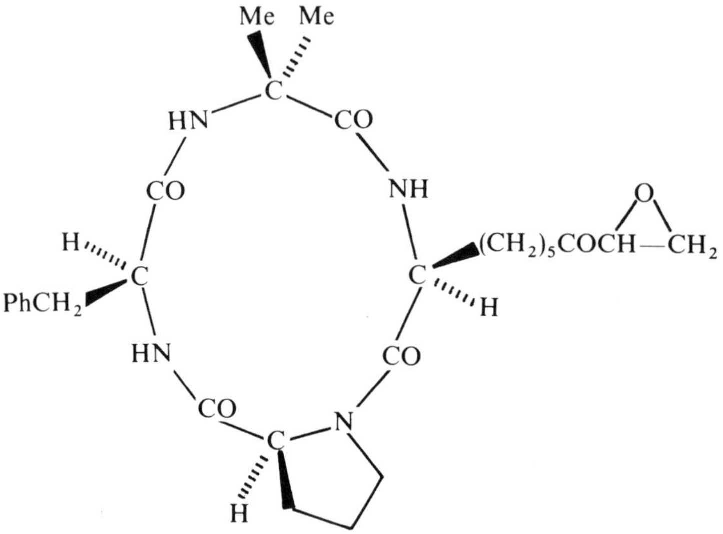

(a) Write out the structures of its constituent amino acids.
(b) Where appropriate, define the stereochemistry at the α-carbons.
(c) Which of the amino acids are DNA encoded?

3. The hormonal peptide oxytocin has the structure shown below. Write out the shorthand notation for its structure.

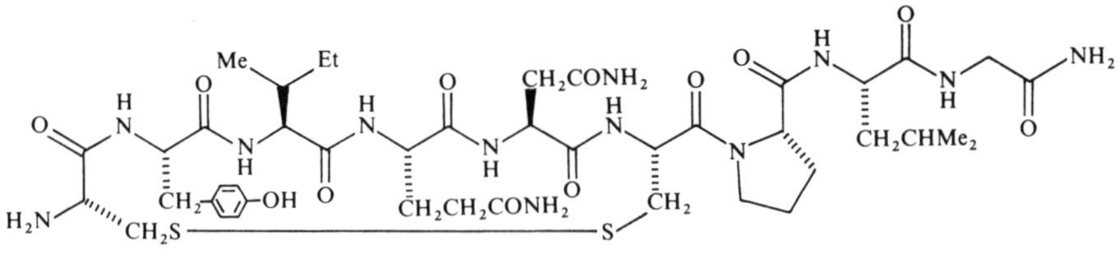

4. (a) For each of the following peptides and amino acid derivatives, draw the full structure from the abbreviation given.

(b) What is the net charge (to the nearest whole number) on each one at pH 1, at pH 6, and at pH 11?

Lys—OEt   Ac—(D)-Tyr   Glu—OMe
                                         |
                                        OMe

(G)   (I)   (K)

Lys
|
Glu
(M)

Phe—(D)-Ala   PhCO—Pro—Asp(OMe)   Lys—Glu

(H)   (J)   (L)

5. (a) What are the products **P–S** from the following reactions of lysine?
   (b) Explain why the three step sequence of reactions gives the N$^\varepsilon$-acetylated derivative of lysine, **T**.

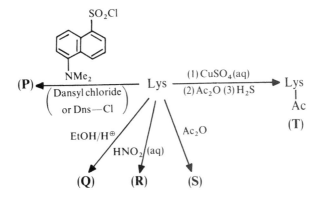

6. The pK_a and pK_b for an amine are derived from equilibria (**1**) and (**2**) respectively:

$$R—NH_2 + H_2O \rightleftharpoons R—\overset{\oplus}{N}H_3 + HO^{\ominus} \qquad (1)$$

$$R—\overset{\oplus}{N}H_3 + H_2O \rightleftharpoons R—NH_2 + H_3O^{\oplus} \qquad (2)$$

$$pK_a + pK_b = 14 \qquad (3)$$

Show that equation (3) is valid, remembering that the equilibrium constant for the dissociation of water ($H_2O \rightleftharpoons H^+ + HO^-$) is 10^{-14}.

CHAPTER 3

Purification and Isolation
of Peptides

At the end of the Chapter 2, we considered some of the physical properties of amino acids. We concluded that these properties would be determined by three main factors:

(i) The net charge at any given pH.
(ii) The pH at which there was no net charge (pI).
(iii) The hydrophilicity or lipophilicity of the side-chains.

In a sense, peptides are just like big amino acids. They contain an amino group at one end, a carboxylic acid group at the other end, and a series of side-chains separated by amide bonds.

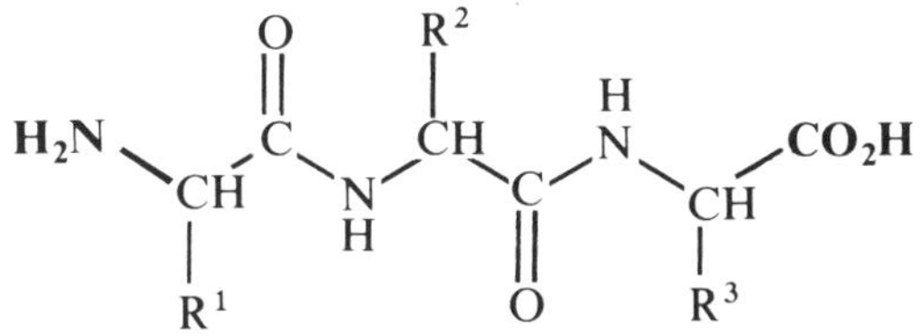

Tripeptide

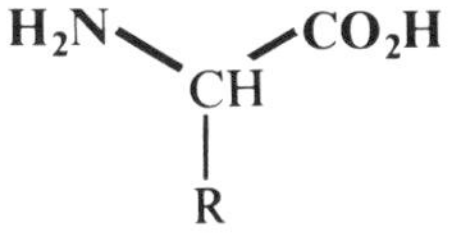

Amino acid

The physical properties of peptides can be roughly determined by simply adding together the effects of each of the side-chains, plus those of the terminal amino and carboxylic acid groups. So peptides, like amino acids, have typical net charges at different pHs, characteristic isoelectric points, and specific affinities for aqueous or non-aqueous phases.

46

There is one further important property that can be used in the purification of peptides—their size. For example, a tripeptide will clearly be much smaller than (say) a decapeptide, and the molecular weights will differ by a factor of about three.

In this chapter, we will look at the techniques that might allow us to purify and finally isolate the peptide PENTIN, which we believe might stimulate the lubrication of cartilage in joints. At this stage, we have simply postulated that the peptide exists; we know nothing about its composition or its physical properties, and we will presumably have to try to isolate it from a large amout of cartilage taken from healthy joints (where the peptide is being generated correctly, we hope). The PENTIN will be just one of thousands of compounds present, so we need to have at our disposal an array of purification techniques that can separate one particular peptide from a mixture of many hundreds of peptides and proteins.

The techniques described below are those most commonly used for the purification of peptides, although many additional methods are known. We will consider size first of all, because it should allow many of the large proteins to be removed from our mixture. Then we will consider how net charge and hydrophilicity might be employed to separate peptides from each other.

1 Dialysis

Semi-permeable membranes are used by many living systems in order to separate molecules of differing molecular size, and this process is called dialysis. For example, the small intestine is designed to allow through only molecules smaller than a particular size (i.e. food that has been properly broken down by the enzymes in the stomach); and kidneys are continually performing dialysis, allowing only water and very small molecules to be excreted as urine.

How do the kidneys and the small intestine carry out their functions? This is achieved simply by having a skin-like structure that has holes in it. The holes or pores are only big enough to let molecules less than a certain size pass through— an extremely simple but efficient idea. Materials that possess this property are

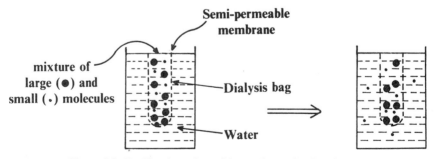

Figure 3.1. Purification of peptides and proteins by dialysis.

called semi-permeable membranes, and the principle is shown in Figure 3.1 for a mixture of a small peptide and a large protein.

Although semi-permeable membranes can be obtained from animal sources, synthetic ones are more usually employed these days. In the example shown above, it is very easy to obtain the small peptide in a pure form. There are, however, two possible problems:

(i) The small peptide will not be partitioned only into the outer vessel: because it is small, it will have free access to all of the solution, and so will be evenly distributed (i.e. will have the same concentration) both inside and outside the semi-permeable membrane. (The large protein, on the other hand, cannot escape from the dialysis bag, and so will be retained within it).

If we want to keep the large protein, then we can just run water through the outer solution. This will wash away the small peptide, which will continue to migrate across the semi-permeable membrane.

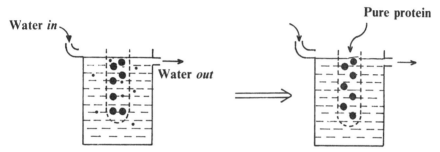

Figure 3.2. Purification of a large molecule by dialysis.

If we want to keep the small peptide, we will need to have a large volume in the outer vessel. When the concentrations are identical on both sides of the semi-permeable membrane, there will be much more of the peptide in the outer vessel than within the bag. We can then collect the outer solution, and evaporate off the solvent; we will then have purified the peptide, without losing more than a fraction of it.

(ii) The other problem is that molecules either pass readily through the membrane, or they are totally impervious to it. In other words, it is easy to separate two molecules of very different size, as in the example shown above. But for a complex mixture with many molecular sizes, the purification is somewhat limited, because there is a simple cut-off-size; all those molecules below the cut-off size pass through the membrane, whilst all those that are too large fail to do so (see Figure 3.3).

So, dialysis is rather a crude method of purification, although it is very frequently used as an early step for the separation of molecules of very different size (e.g. peptides from proteins), and it is quite easy to handle large quantities of material using dialysis. If you have a range of semi-permeable membranes, then

48

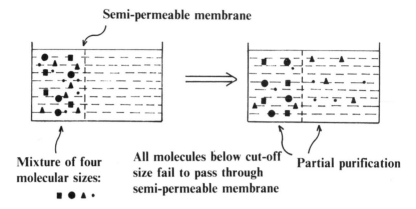

Semi-permeable membrane

Mixture of four molecular sizes:
■ ● ▲ ·

All molecules below cut-off size fail to pass through semi-permeable membrane

Partial purification

Figure 3.3. Example of limitation of purification by dialysis.

you can fractionate a mixture of peptides and proteins by molecular size, but a much simpler method is to use gel filtration.

> Dialysis involves the use of semi-permeable membranes. Whilst small molecules can pass through the pores in the membrane, large molecules are too big, and are retained within the dialysis bag.

2   Gel Filtration

This technique also separates compounds on the basis of their size, and uses a polymeric support that has a relatively open structure. The mixture of peptides and proteins is loaded onto the top of a column of the polymer, and is eluted with a suitable solvent—usually buffered water. The principle is that only small molecules can enter the open structure (or pores) in the polymer, and so they are eluted rather slowly; the larger molecules come off the column first, as they are unable to enter the pores.

It is easy to see the principle of gel filtration if we consider a simple mixture of a large peptide and a small peptide.

The polymer has been chosen to have an intermediate pore size.

The large peptide is eluted first, because there is only a very small volume available to it. For example, if the volume of the empty column is $100 \, cm^3$, and the space occupied by the support is $90 \, cm^3$, then the large peptide will be eluted after $10 \, cm^3$ of solvent have been run through.

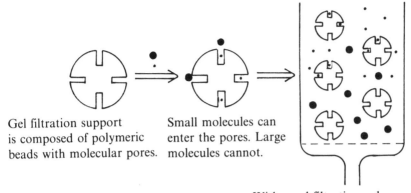

Gel filtration support is composed of polymeric beads with molecular pores.

Small molecules can enter the pores. Large molecules cannot.

With a gel filtration column, the smaller molecules (·) experience a larger column size than the large molecules (●). The large molecules are therefore eluted first.

Figure 3.4. Separation of a large peptide and a small peptide by gel filtration.

On the other hand, it might be discovered that the support absorbs 40 cm³ of solvent into its pores. This would mean that an additional 40 cm³ of column would be available to the small peptide, and it would not be eluted until 50 cm³ of solvent had been run through.

For a complex mixture of peptides, it should be clear that a single pore size would mean that almost all molecules would experience either a small (10 cm³) column volume, or a large (50 cm³) column volume, and you would simply get two big elution peaks after 10 cm³ and after 50 cm³. By having a **range** of pore sizes, very small molecules would enter all of the pores, medium sized molecules would enter some of them, and large molecules would enter none of them. This would ensure that a good overall separation would be obtained. For molecules of intermediate size, diffusion factors also play an important role, and retention times are approximately a linear function of log(MW).

If a really broad range of pore sizes is employed, the peaks tend to be rather poorly resolved. The compromise is to choose a polymer that has a fairly small range of pore sizes that are close to the molecular size that you are expecting. Even so, really well-resolved peaks are rarely seen when peptides are purified using gel filtration—but you can achieve quite dramatic improvement in purity very easily, and get an idea of the molecular weight of the peptide at the same time. Other techniques are usually needed in order to isolate a peptide in a really pure state.

50

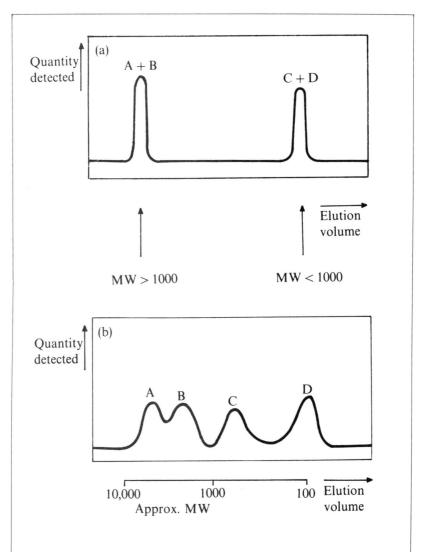

Figure 3.5. Purification by gel filtration. A mixture of four peptides, with MW 5000 (A), 3000 (B), 500 (C), and 100 (D):
(a) A single pore size gel filtration column. With the pore size corresponding to a MW of about $1000 (\simeq 12 \text{ Å})$, A + B would be eluted together, as would C + D (cf. the cut-off size in dialysis).
(b) With a *range* of pore size, *averaging* about 12 Å, all four peptides would be separated; this is a fairly typical gel filtration elution profile, and gives an indication of the molecular weights of the peptides.

Gel filtration employs a porous polymeric support. Large molecules which cannot enter the pores are eluted rapidly, whilst smaller molecules have longer retention times.

3 Ion Exchange Chromatography

Like their constituent amino acids, peptides can possess a net charge (which may be positive or negative) at any pH, and this net charge is greatly influenced by the nature of the side-chains. If a solution of a mixture of peptides is passed down a column that contains a charged polymeric support, then they will elute at different rates, dependent mainly on their net charge. This is the principle of ion exchange chromatography, and typical ion exchange resins can be either positively or negatively charged:

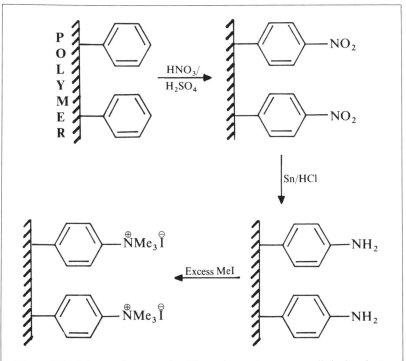

Figure 3.6. Anion exchange resin. The polymer (e.g. a cross-linked polystyrene) is converted into an ammonium derivative. Initially, a counter-ion (such as I⁻, in this case) is present. Other anions would have an affinity for the positive charge, and could displace (or exchange with) the first ion—hence 'anion exchange chromatography'.

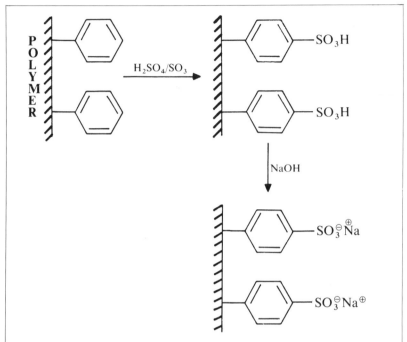

Figure 3.7. Cation exchange resin. Functionalisation of a polymer with a negatively charged group such as sulphonate allows 'cation exchange chromatography' to be performed.

Negatively charged polymeric supports are called cation exchange resins, because cationic species have a high affinity for them. Conversely, positively charged polymeric supports are called anion exchange resins.

Quite clearly, negatively charged peptides would have a high affinity for an **anion exchange resin**, whilst positively charged peptides would have little affinity for such a polymer, and so would be eluted rapidly. In fact, negatively charged peptides might bind so strongly that they would fail to be eluted at all. To make sure that everything is eluted within a reasonable time span, the eluant is often varied with time, and this frequently improves the resolution of the components as well:

(i) The pH of the eluant can be changed as time progresses. This will affect the overall net charge on the peptide, and at the isoelectric point the peptide would be expected to elute quite rapidly.

(ii) The polarity of the solvent can be increased as time progresses, so the polar peptides have an increasing affinity for the aqueous phase. This is achieved by increasing the ionic strength of the solvent (by increasing the concentration of, say, sodium chloride), and it is a fairly subtle way of separating quite similar molecules.

Below are two examples of mixtures of peptides that could be readily separated by ion exchange chromatography, using a cation exchange resin, and a linear variation in eluant.

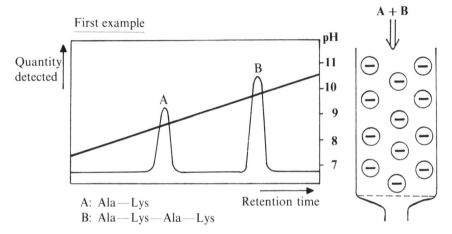

First example

A: Ala—Lys
B: Ala—Lys—Ala—Lys

Below pH9, both A and B would be eluted slowly from a cation exchange column, because they both carry a net positive charge. Peptide A would be eluted first, because it would be less positively charged; by using a pH gradient, both peptides are eluted reasonably rapidly, and will emerge at a pH close to their iso-electric points (pI).

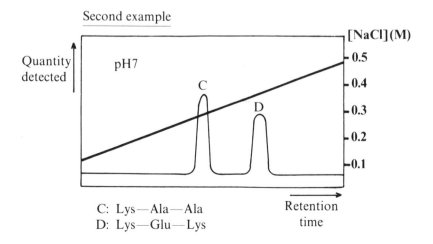

Second example

C: Lys—Ala—Ala
D: Lys—Glu—Lys

Between pH 3 and pH 10, peptides C and D will carry very similar net charges (e.g. approx. + 1 at pH 7); they might therefore be difficult to separate on an ion exchange column, even with a pH gradient. However, peptide D contains more polar groups, and would therefore be eluted more slowly from the highly charged resin. The different affinities for a polar solvent can be exaggerated by running an ionic strength gradient. By steadily increasing the polarity of the solvent, C and D are efficiently resolved, and the retention times are kept reasonably short.

Ion exchange chromatography utilises the net charge carried by most peptides. When a mixture of peptides is passed down a negatively charged support (cation exchange resin), positively charged peptides are eluted slowly, whilst negatively charged ones are eluted rapidly. The converse is true with anion exchange resins.

---

4 Electrophoresis

In electrophoresis, a mixture of peptides is loaded onto a solid support. The support must be permeated with a liquid electrolyte, so that a potential difference can be applied across it. When the current is switched on, positively charged peptides are attracted to the cathode, and so migrate towards that electrode. Conversely, negatively charged peptides migrate towards the anode.

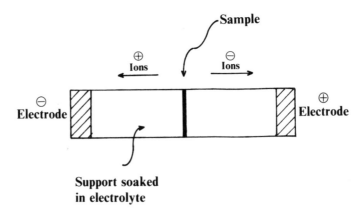

Figure 3.8. Principle of electrophoresis.

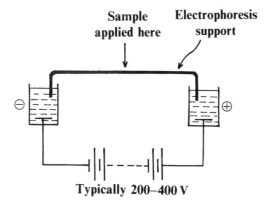

Figure 3.9. Paper electrophoresis.

The simplest support is a strip of paper, and the peptides are then applied near to the centre of the strip. The rate of migration of different peptides will depend on their net charge, and on their resistance to motion. So a small, highly charged peptide would migrate much faster than one that was large, or that carried only a small net charge.

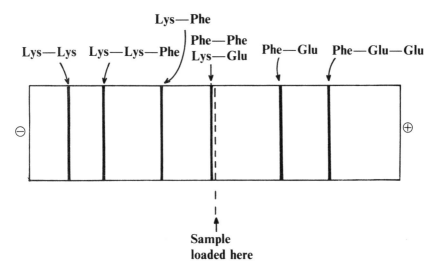

Typical electrophoretogram from a mixture of di- and tri-peptides, run at pH7. The dipeptides Phe—Phe and Lys—Glu are electrically neutral, and stay close to the origin. The higher the net charge, the further the peptides migrate (e.g. Lys—Lys > Lys—Phe), but an increase in size reduces mobility (e.g. Lys—Lys—Phe migrates less far than Lys—Lys).

Paper electrophoresis can be an extremely effective method of purifying small quantities of peptides. There are a couple of variations that broaden the applicability of the technique considerably.

4.1 SDS-PAGE Electrophoresis

Instead of using paper as the support, a polymer such as polyacrylamide can be prepared as a gel (with about the same consistency as jelly), and the mixture of peptides can then be loaded at one end. This is called polyacrylamide gel electrophoresis, usually abbreviated to PAGE.

Radical mechanisms: For clarity, only one of the one-electron single-headed arrows is shown when double bonds react by homolytic cleavage.

Polyacrylamide gels are formed by the radical-initiated polymerisation of acrylamide. Cross-linking agents [e.g. $CH_2(NHCOCH{=}CH_2)_2$] are added, to give solidity to the gel.

You can run PAGE in exactly the same way as paper electrophoresis. However, if it is carried out in the presence of sodium dodecylsulphate (SDS), then three things happen.

$$CH_3(CH_2)_{10}CH_2 - \overset{\displaystyle O}{\underset{\displaystyle O^{\ominus}Na^{\oplus}}{\overset{\|}{S}}} = O$$

Sodium dodecylsulphate (SDS)

(i) The SDS is a detergent (or chaotropic agent). Its ionic sulphate group and lipophilic alkyl chain cause peptides and proteins to unfold. This process is called denaturation.

(ii) The SDS surrounds the peptide or protein, giving all such molecules a net negative charge. This means that all the peptides and proteins in the mixture migrate towards the anode.

(iii) Finally, and most importantly, the SDS causes the peptides and proteins to migrate **as a function of their molecular weight alone.** Large proteins migrate slowly (because they are bulky, despite their high negative net charge), whilst smaller molecules migrate more rapidly.

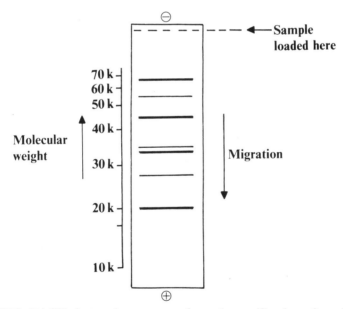

Typical SDS–PAGE electrophoretogram, from the purification of a mixture of proteins. The molecular weight can be deduced from the distance migrated, usually by comparison with known standards. The molecular weight scale is non-linear, and is usually quoted in kilo-daltons (abbreviated to kD or k); $1 kD \equiv$ molecular weight of 1000.

Thus SDS-PAGE is an excellent method for separating peptides and proteins on the basis of molecular weight alone, and it allows the approximate molecular weight to be determined as well. However, it is quite hard to detect small peptides in a polyacrylamide gel, and paper electrophoresis is usually preferred for these compounds, particularly as isoelectric focusing can also be readily applied.

4.2 Isoelectric Focusing

If two peptides, of similar molecular weight, differ only slightly in their net charge at a given pH, then it is often very difficult to separate them using simple electrophoresis. Furthermore, the longer you run the electrophoretogram in order to separate the molecules, the more diffuse will the components become. However, if the electrophoresis is conducted across a pH gradient, then the longer the electrophoretogram is run, the sharper the bands become. Because of this, the technique is known as isoelectric focusing.

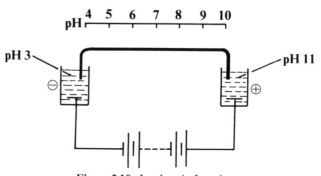

Figure 3.10. Isoelectric focusing.

For example, let us consider two compounds that have isoelectric points at pH 8.0 (peptide A) and pH 8.1 (peptide B). So, if the centre of the electrophoretogram was at pH 7, they would both start to migrate to the cathode (as they both hold a net positive charge), and they would start to diffuse as they travelled. The molecules of peptide A that arrived first at the position at which the pH was 8.0 would stop moving, as they would have no net charge; any molecules of peptide A that got left behind would carry on moving until they reached the point where the pH was 8.0. Peptide B would continue migrating to the cathode, and would eventually start concentrating at the point where the pH was 8.1.

Another way to visualise this focusing effect is to assume that peptide B would be rather spread out when it arrived at the point where the pH was 8.1. But if the electric current was maintained, then those molecules that had overrun would start to go backwards, because they would hold a slight negative charge. Conversely, those molecules that got left behind would continue to catch up, because they would still carry a slight positive charge. So, as long as the potential difference was maintained, the two peptides would form progressively sharper and more distinct bands.

> Electrophoresis separates peptides by applying a potential difference across a support holding the sample. Peptides migrate because of their net charge, with small highly charged peptides moving fastest.

5 High Performance Liquid Chromatography

The key feature of high performance liquid chromatography (HPLC) is that it is conducted under pressure; the term 'high performance' has arisen because of the quality of the separations that can be achieved.

In order to obtain a good separation on any chromatography column, it should ideally be very long and evenly packed. Such columns would take many hours to run, even if they could be packed sufficiently well. If the effective surface area of the material in the column is very large, then this has the same effect as having a very long column; but in practice it would similarly take a long time for the solvent to elute. However, if the solvent is passed through the column under pressure, then a reasonable flow rate can be achieved, and this is the principle of HPLC.

Modern supports have been developed that are composed of extremely fine particles; this increases the effective surface area of the support, so that columns of (say) 20 cm length are as efficient as old-fashioned columns of over 1 m. When run at pressures of 10–100 atm, elution occurs very rapidly, and a typical HPLC run would take 10–30 minutes. So reliable is this technique that a major criterion for the purity of a peptide is that it gives a single peak on an analytical HPLC column. Analytical HPLC columns can efficiently analyse only small amounts of peptide sample (< 1 mg), although preparative HPLC columns are now available (at a price!) that can purify up to a gram of peptide.

Up until now, there has been no mention of the **type of support** in the HPLC columns. This is because any type of stationary phase can be used, as long as it is possible to prepare it with a large effective surface area. The following types of column are available:

(i) Ion exchange (very widely used).
(ii) Gel filtration (good on small quantities).
(iii) Silica (excellent with non-polar organic molecules, but not so good for peptides).
(iv) Reversed phase.

Reversed phase HPLC is a particularly effective method of purifying peptides. Silica (SiO_2) or alumina (Al_2O_3) are typical supports for column chromatography in standard organic chemistry; however, they are less good supports for the purification of peptides, which tend to remain on the baseline unless very polar (slow-moving) eluants are used. In reversed phase chromatography, lipophilic alkyl groups are attached to the support. So a sample running down the column in a polar solvent is continually being partitioned between a 'hydrocarbon' stationary phase, and a more polar mobile phase. This means that very polar peptides tend to be eluted rapidly, whereas less polar, lipophilic ones are eluted slowly.

> Reversed phase chromatography uses a non-polar stationary phase, and a polar mobile phase. (Normal adsorption chromatography is the other way round!) Hydrophilic peptides are eluted rapidly, whereas lipophilic ones have longer retention times.

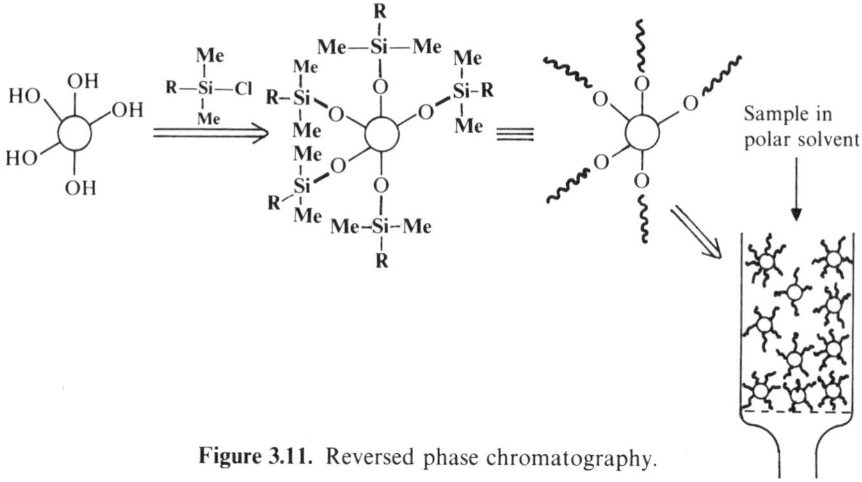

Figure 3.11. Reversed phase chromatography.

> **Reversed phase chromatography.** The support consists of fine beads (e.g. $< 10\,\mu m$ diameter) of silica, which always contain Si—OH groups on the surface; these are silylated by treatment with RMe_2SiCl. The R group is a hydrocarbon chain—the longer the chain, the more lipophilic the column.

So reversed phase chromatography exploits differences in the hydrophilicity of peptides in a mixture, and aqueous eluants are usually employed. If the polarity of the solvent is decreased, then the less polar peptides start to be eluted more easily. In practice, it is often most convenient to run reversed phase HPLC columns with an increasing concentration of a less polar organic solvent as time progresses; gradients involving ethanonitrile (MeCN) or simple alcohols (MeOH or Me_2CHOH) are typical. Of course, the pH of the run (which affects the ionisation/polarity of the peptides) is also important, but a constant pH close to neutrality is usually employed.

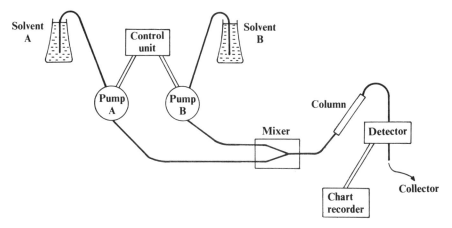

Figure 3.12. Typical arrangement for HPLC.

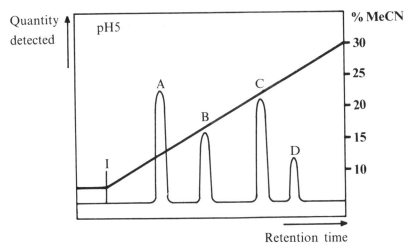

Typical HPLC elution profile, for the separation of a mixture of four peptides. The sample was injected (I), at which point the solvent gradient was started. The most hydrophilic peptide was A, and the most lipophilic was D.

HPLC is a technique that employs high surface area supports that can withstand quite high pressures. The support can be of several kinds, which allow properties of size (gel filtration), net charge (ion exchange), or hydrophilicity/lipophilicity (reversed phase) to be exploited.

6 Detection

Up until now, we have been considering the different methods that might be used for the separation of peptides, without worrying about how they might be actually detected during the purification procedure. We will now consider three different types of detection that can be used.

6.1 Non-destructive Methods

This essentially means using some physical or spectroscopic property, and allows peptides to be recovered intact. A number of methods are available:

6.1.1 *UV Absorption.* The most commonly used method for detecting peptides eluting from a chromatography column is to monitor the UV absorption of the eluant. For peptides, there is always a λ_{max} at about 215 nm ($\pi \rightarrow \pi^*$ transition for the amide carbonyl); it is usually more convenient to monitor a wavelength on the shoulder of this absorption, because most eluants absorb significantly at 210 nm; 230 or 254 nm are commonly used. If the peptide contains aromatic side-chains (e.g. Phe or Tyr), then absorption at 280 nm can be monitored instead.

6.1.2 *Refractive Index.* When a compound is present in an eluant, the refractive index (RI) of the solvent changes. Interference refractometers are extremely sensitive, but (unlike UV absorption) cannot be easily used with solvent gradients (because the RI of the solvent itself will be changing), so their use is a little limited.

6.1.3 *Optical Rotation* (OR). Because peptides contain chiral amino acids, they are able to rotate polarised light. Very sensitive laser OR detectors are becoming available, and this could become an important method of detection.

6.2 Destructive Chemical Methods

Non-destructive methods are obviously preferred for the detection of peptides whenever possible, and are usually employed for the analysis of products from chromatography columns. Sometimes these methods are insufficiently sensitive; furthermore, they are unsuitable for the analysis of electrophoretograms. Therefore, a number of special chemical tests have been devised instead:

  (i) **Ninhydrin** is often used to aid detection; it gives a distinctive purple colour with peptides containing a free primary amino group—although gentle heating is often required (cf. purple colour with amino acids, see pages 25–26). This is often used to develop paper electrophoretograms.
 (ii) **Fluorescamine** reacts with any primary amino group. The resulting derivatives absorb light at 390 nm, and re-emit it at 475 nm; by monitoring this **fluorescence**, it is possible to detect picomole quantities (10^{-12} mole) of peptides or amino acids.

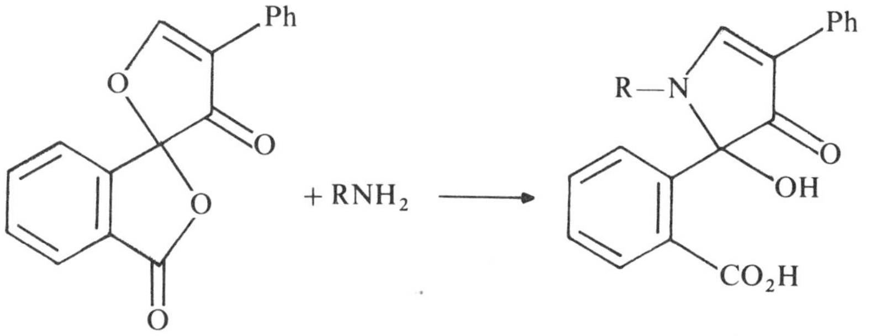

Fluorescamine

Primary amine adduct
Absorbs light at **390 nm**
Re-emits light at **475 nm**

(iii) **O-Phthalaldehyde** (OPA) is another extremely sensitive reagent. If a primary amine is reacted with OPA in the presence of mercaptoethanol $(SH—CH_2CH_2—OH)$, then the product will fluoresce, absorbing light at 340 nm and re-emitting it at 450 nm (cf. fluorescamine derivatives above). Again, it is possible to detect minute quantities of peptides or amino acids using this reagent.

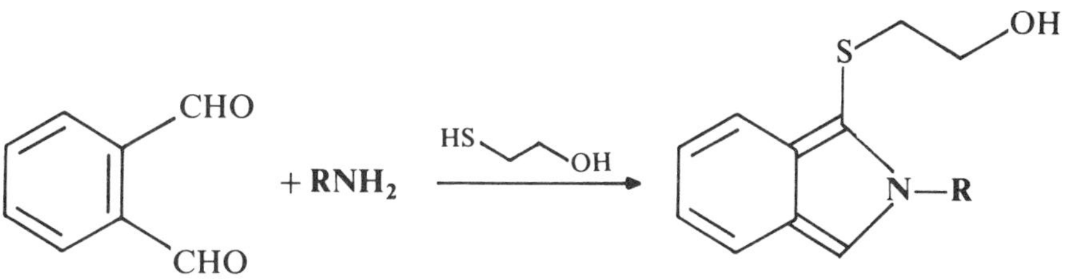

OPA

Primary amine adduct
Absorbs light at **340 nm**
Re-emits light at **475 nm**

(iv) **Coumassie** stain is a reagent which gives a blue colour with proteins, and is convenient for developing gel electrophoretograms; peptides, however, show up only poorly.

Destructive methods for detecting peptides are acceptable, provided that you are simply **checking the purity** using a small amount of material. However, if (for example) you were using paper electrophoresis to actually **purify** a peptide, then it would be useless to have to destroy it in order to visualise it. In this case, it is normal to remove and visualise a very thin strip of the paper, in order to determine how far the peptide has migrated:

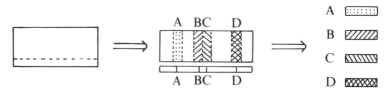

To isolate peptides from an electrophoretogram, a strip can be cut from one edge. The peptides can then be visualised (e.g. spray with ninhydrin), and their positions on the electrophoretogram deduced. Bands containing the peptides can then be cut off, and the peptides can be washed off the bands and isolated.

6.3 Biochemical Methods

Although most of the methods described above can allow peptides to be detected at the μg level, this is not always sensitive enough. Moreover, if large amounts of other peptides and proteins are present, the techniques do not **select** for the particular peptide that you are trying to isolate. To get round this problem, a number of extremely sensitive biochemical tests have been devised, which can allow **minute quantities** (often $< 10^{-15}$ mole) of **specific peptides** to be detected.

Biochemical methods of detection have been crucial to the isolation of many peptides, and there is insufficient space for this important area to be discussed in more detail. But the biochemical tests for specific peptides are usually based on the biological property that you hope the peptide will exhibit; for example, it is easy to see that a peptide that is thought might cause blood to clot (e.g. Factor VIII, which is given to haemophiliacs) could be tested directly on blood samples. However, most biochemical tests rely on the use of extremely specific antibodies which can bind to the peptide being studied; a variety of chemical or radio-labelling tests can then be used to determine whether the antibody has become complexed to the peptide.

7 Summary

How could we purify and isolate a new, biologically active peptide? For our peptide PENTIN, which we believe might halt the progression of damage in arthritic joints, we would probably start by extracting the water-soluble components from healthy joints. This might give us several hundred grams (or even a few kilos) of material that might contain PENTIN **plus hundreds of other peptides and proteins**.

Quite early on, we would undoubtedly have to devise a biological test for PENTIN; in other words, we would have to develop some analytical method that would reveal which fractions from the purification steps actually displayed the desired biological property. For example, we might be able to measure the rate at which a piece of cartilage (in a petri dish) was destroyed by the fluid extracted from arthritic joints—a simple staining technique could be used to detect the chemicals released from within a cell (e.g. proteins/DNA/RNA/lipids) when it is destroyed; if fractions containing PENTIN were added to the assay system, then cell destruction would be reduced, and the staining would be slower (or less intense). This would allow us to identify those fractions containing PENTIN.

There is no single purification method that would allow us to isolate appreciable quantities of PENTIN. Initially we would have to utilise methods that could cope with large quantities of material, and this would give rather crude purifications. But as the amount of sample became smaller, the higher-resolution chromatographic techniques would become viable. So a typical combination of purification steps might be as follows:

(i) Dialysis, to remove proteins.
(ii) Gel filtration, giving significant purification, and an approximate molecular weight.
(iii) Ion exchange chromatography, because relatively large amounts of material can be readily handled. At this stage, non-destructive detection methods might allow PENTIN to be visualised.
(vi) Reversed phase HPLC, which is excellent for purifying small amounts of peptides.
(v) Paper electrophoresis, which could be used to check the purity of the peptide and, if necessary, for the small-scale isolation of really pure PENTIN.

At the end of these purification steps, we will assume that we have isolated 10 mg of absolutely pure PENTIN. From the gel filtration results, we know that its molecular weight is about 500. The next stage is to determine which amino acid residues are present in PENTIN, and this is discussed in Chapter 4.

Further Reading

Laboratory Techniques in Biochemistry and Molecular Biology, Volume 9, *Sequencing Peptides and Proteins*, 2nd Edition, G. Allen, Chapter 4, pp. 105–76, Elsevier, 1989. Covers all the main methods for separating peptides, with a strong practical bias.
The Peptides; Analysis, Synthesis, Biology, Volume 4, *Modern Techniques of Conformational, Structural, and Configurational Analysis*, (E. Gross and J. Meienhofer, Eds.), Chapter 4, pp. 185–216. Looks at the use of HPLC for the purification of very small quantities of peptides, covering a range of columns and detection methods.

Questions

1. After partial purification, the extracts from a gland that controls the growth of rats was found to contain six peptides, whose structures were eventually

deduced from sequencing experiments:

A Lys—Phe
B Lys—Glu—Phe
C Lys—Lys—Phe
D Ala—Leu—Val—Ala—Phe
E Ala—Asp—Val—Ala—Phe—Gly
F Cys—Asp—Glu—Ala—Phe—Cys

State the approximate order in which you would expect them to migrate or elute using the following purification techniques:

(a) Gel filtration
(b) Ion exchange chromatography (R—N Me$_3^+$ support, at pH 7)
(c) Reversed phase HPLC (at pH 5)
(d) Electrophoresis (at pH 7)

How might a combination of just two of the methods allow the separation of all six peptides to be accomplished?

2. (a) Starting from a suitably cross-linked polystyrene, how might the following cation exchange resin be prepared? (*Hint*: one approach might be to use chloromethylation—see page 141)

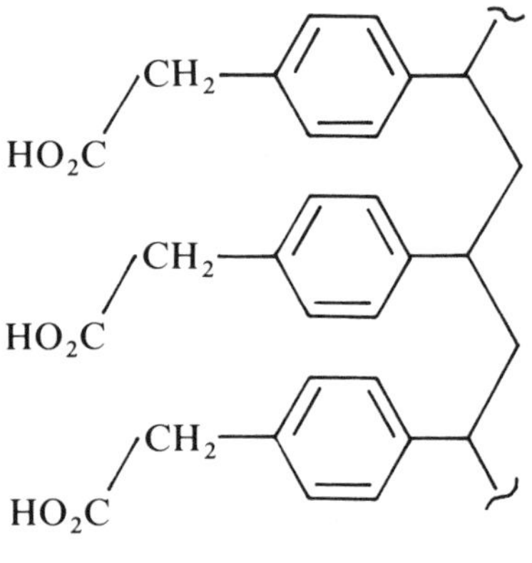

(b) In what order would the peptides **A–F** from question 1 elute from such an ion exchange column, at pH 6?

3. The enzyme chymotrypsin cleaves the peptide bond on the carboxy side of aromatic residues (e.g. Phe, Trp, Tyr); trypsin cleaves them on the carboxy side of basic residues (e.g. Arg, Lys). Suggest the simplest technique for separating the fragments resulting from the treatment of **X** with:

(a) chymotrypsin;
(b) trypsin.

Pro—Arg—Val—Phe—(D)-Pro—Leu—Ala—Trp—Lys—Asp—Asp—Gly—Gly

X

CHAPTER 4

Amino Acid Analysis

At the end of the last chapter, we had managed to isolate our peptide PENTIN, and had deduced that it had a molecular weight of about 500. Our next problem is to determine which amino acids are contained in PENTIN. Fortunately we can assume (initially, at least) that PENTIN is a simple DNA encoded peptide, and so it must be composed only of the 20 commonly occurring L-amino acids.

In order to identify the amino acids present, we must first completely hydrolyse the amide bonds, to give the free amino acids; then we must separate, identify, and quantify them.

1 Total Hydrolysis of Peptides

Although amide bonds are relatively unreactive, they can be hydrolysed under strongly acidic or alkaline conditions, or by the use of specific enzymes.

1.1 Acidic Hydrolysis

This is the most widely used method of totally hydrolysing peptides for amino acid analysis. The standard conditions are:

6M HCl(aq)
110 °C
24 hours

The conditions for acidic hydrolysis are quite harsh, which emphasises the stability of the peptide bond. The reaction is usually carried out in a sealed glass tube, which is thoroughly degassed first.

At the end of reaction, the water is removed under high vacuum, and the amino acids can then be taken up in a suitable solvent (usually water) ready for analysis. The mechanism for the hydrolysis is as follows:

68

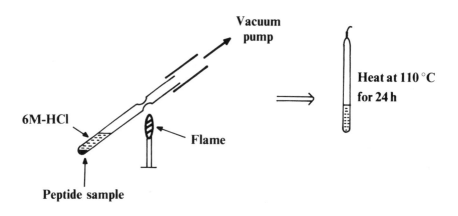

Mechanism for acid-catalysed hydrolysis of amide bonds.

Total acidic hydrolysis of peptides. The sample is sealed *in vacuo* with 6M-HCl. After heating for 24 h at 110 °C, the tube is broken open, and the solvent evaporated off, leaving the constituent amino acids of the sample.

One of the problems with acidic hydrolysis is that some of the amino acids can be destroyed under the reaction conditions.

(i) Ser and Thr. Serine and threonine are slowly dehydrated to give the corresponding alkenes, but the losses are not usually serious.

(ii) Met, Cys and Cys$_2$. The sulphur-containing amino acids are readily oxidised under acidic conditions at 110 °C, so oxygen must be rigorously excluded. The degassing procedure removes most of the dissolved air, but the reaction mixture can also be thoroughly flushed with nitrogen.

(iii) Trp. Tryptophan is completely destroyed under conditions of strong acidic hydrolysis, giving a mixture of products. Therefore, it is not possible to analyse for tryptophan after total acidic hydrolysis.

(iv) Any amide side-chains will also be hydrolysed to the corresponding acid. So asparagine or glutamine residues would be converted to aspartic acid or glutamic acid respectively.

1.2 Alkaline Hydrolysis

Amide bonds can be efficiently cleaved by treatment of a peptide with 2M NaOH(aq) at 100 °C. The reaction proceeds because of the nucleophilicity of the hydroxide ion:

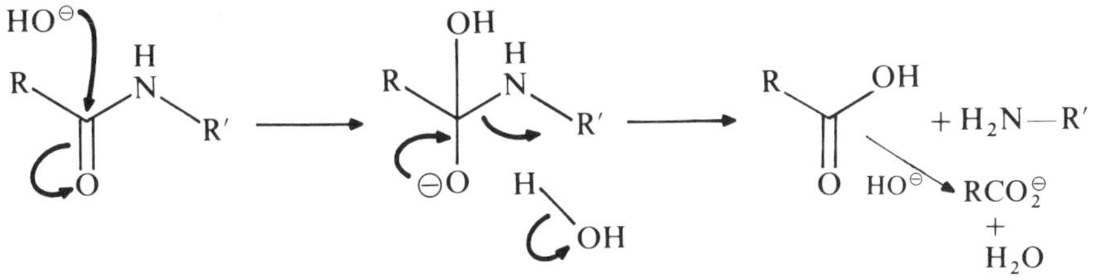

Mechanism for base-catalysed hydrolysis of amide bonds.

Alkaline hydrolysis results in the complete destruction of arginine, cystine, serine, and threonine. However, **tryptophan** is **not** destroyed under these conditions; if its presence is suspected, then alkaline hydrolysis may well be used to liberate the constituent amino acids from a peptide. However, in general, acidic hydrolysis destroys fewer of the amino acids, and gives more reliable results.

1.3 Enzymic Hydrolysis

In order to break down food, the gut contains a range of enzymes that can catalyse the cleavage of peptide bonds—they are known as peptidases. The aminopeptidases are particularly rapid and efficient in their hydrolysis of peptide bonds, and cleave off amino acids one residue at a time starting from the N-terminus.

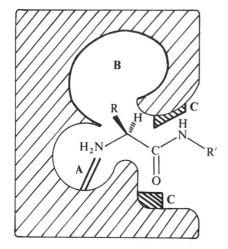

Schematic drawing of an aminopeptidase. The peptide is bound by the terminal amino group (A), and a pocket in the enzyme can accommodate the side-chain (B). This aligns the peptide so that the catalytic sites in the enzyme (C) can cleave the amide bond. Peptides of the D-configuration, or those not possessing a free α-amino group, are unable to bind to the enzyme.

Only a tiny amount of the enzyme is required to catalyse the total hydrolysis of most peptides, under very mild conditions—typically 37 °C at pH 7. However, there are two specific residues that block any further hydrolysis by aminopeptidases:

(i) Proline. This secondary amino acid residue, in which the side-chain is joined on to the α-amino group, cannot adopt the correct conformation for the enzyme to catalyse cleavage of the peptide bond.

(ii) D-Amino acids. The side-chains of these residues are unable to fit into the catalytic pocket of the enzyme. This means that the peptide fails to bind to the enzyme, and peptide bond cleavage cannot take place.

In a similar way, carboxypeptidases cleave peptide bonds sequentially from the carboxy terminus. However, the rate of hydrolysis is generally much slower than for aminopeptidases, and carboxypeptidases are more often used as part of sequencing studies (see Chapter 5).

So aminopeptidases will 'rapidly' cleave off amino acids sequentially, until they encounter a proline residue or a D-amino acid, at which point hydrolysis will stop. They are particularly useful for determining whether D-amino acids are present in a peptide, but the general applicability of this method for the total hydrolysis of peptides is rather limited.

2 Separation of Amino Acids

Having carried out the total hydrolysis of a peptide, the next problem is to separate the liberated amino acids. We came across several possible methods in Chapter 3, when we were considering the purification of peptides.

2.1 Ion Exchange Chromatography

This is usually an extremely efficient method of separating mixtures of amino acids. Either negatively or positively charged supports could be used, but in practice the negatively charged sulphonate derivatives ($ArSO_3^-$) are used almost exclusively. These are called cation exchange resins, because positively charged cations are attracted to the support.

In Chapter 2, we deduced that amino acids are present as zwitterions at pH 7:

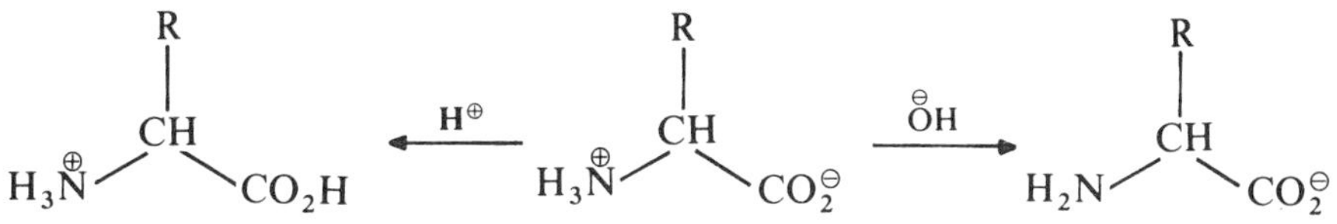

pH7

Because of the cationic ammonium ion, all of the amino acids will have some affinity for the cation exchange columns. However, the **relative retention times** will be largely determined by the **net charge** on the amino acids. The side-chain (R) may have a simple inductive effect on the degree of ionisation at different pHs, or it may itself possess an ionisable group. To see how this might affect the retention times for different amino acids, we will consider again the four amino acids that we looked at in detail in Chapter 2 (Gly, Phe, Glu, and Lys); the net charge at pH 7 and at pH 2 can be found by simply looking at the graph on page 38.

---

pH 7

Glycine (Gly) simply has H for the side-chain, and will have almost no net charge.

Phenylalanine (Phe) has a CH_2Ph side-chain which exerts a $-I$ effect; even so, at pH 7, there will be virtually no net charge.

Glutamic acid (Glu) has an additional carboxylic acid group on the side-chain, and overall will possess a negative charge.

Lysine (Lys) has an additional amino group on the side-chain, and overall will possess a positive charge.

The amino acids would therefore be eluted from an $ArSO_3^-$ cation exchange column in the following order, at pH 7:

Glu   (net charge -1)
Gly and Phe (net charge 0)
Lys   (net charge $+1$)

We would expect that Gly and Phe (and indeed any other amino acids with non-ionisable side-chains) would elute after very similar times at pH 7.

---

pH 2

The amino acids will all have non-integral net charges at pH 2 (see graph on page 38), because this is close to the pK_a for the carboxylic acid group:

Glycine (Gly) $+0.7$
Phenylalanine (Phe) $+0.4$
Glutamic acid (Glu) $+0.6$
Lysine (Lys) $+1.6$

At pH 2, the amino acids would therefore be eluted from an $ArSO_3^-$ cation exchange column in the following order:

Phe
Glu
Gly
Lys

This time, all four of the amino acids would have different retention times, and could be readily separated. However, because they all carry considerable net positive charges, they would be eluted very slowly from the cation exchange column—lysine would probably take days to come off.

In order to separate all of the amino acids in a reasonable time, pHs between 3 and 10 are commonly used. In fact, the pH is often increased whilst running the column, so that acidic, then neutral, then basic amino acids are eluted reasonably rapidly. If the amino acids are ones encoded by DNA, then they can be identified by comparison with the known retention times for the common amino acids.

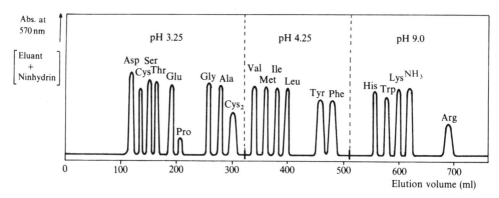

Cation exchange chromatography for the separation of standard DNA encoded amino acids. Equi-molar quantities of all the amino acids were used; note that not all the amino acids are detected with the same efficiency—standards are therefore needed not only to determine retention times, but also in order to allow quantitative analysis. Detection is usually determined by the absorption at 570 nm of the ninhydrin derivatives. (Small amounts of ammonia, often generated during the hydrolysis of peptides, are frequently present too.)

So, if a small amount of a peptide was subjected to total acidic hydrolysis, and then passed down a cation exchange column, the constituent amino acids could be identified by comparison with the standards.

For example:

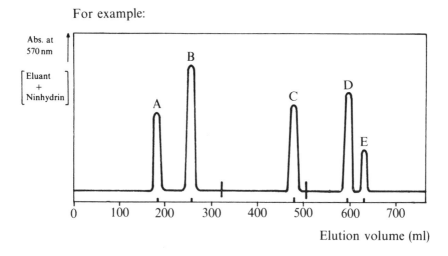

Amino acid analysis after the total acidic hydrolysis of a peptide. By comparison with the standards (page 72), the peaks can be identified as follows:

A: Glu   B: Gly   C: Phe   D: Lys   E: NH_3

There still remains one question to be answered: how are the amino acids actually detected? The normal method is to react the eluant with ninhydrin, immediately before passing it through a spectrophotometric cell. As we discussed in Chapter 2, a purple colour is produced (see pages 25–26), and the absorption at λ_{max} (570 nm) can be automatically monitored, to give a measure of the amount of amino acid present. There are just two things that one needs to watch out for:

(i) Proline, which is a secondary amine, gives only a pale yellow colour with ninhydrin, and this has to be detected separately at a shorter wavelength.

(ii) Not all amino acids react equally well with ninhydrin, and the amount of colour that they give varies accordingly. Therefore, in order to obtain reliable results, it is usual to run a standard equimolar mixture of the DNA encoded amino acids through the column. This does not create any extra work, because the standards would normally be run anyway, in order to check the expected retention times.

The minimum amount of amino acid that can be readily quantified using the ninhydrin test is about 1 μg. This means that you only need about 0.01 mg of a pure decapeptide in order to be able to determine its amino acid composition. For even smaller quantities of amino acids, the fluorescence techniques are required—see fluorescamine and OPA derivatives on pages 62–63. The use of ion exchange columns in parallel with ninhydrin detection methods is so well understood, and the results are so reproducible, that it has become the most widely used method for separating and analysing mixtures of amino acids. There are several alternative methods, however, which are gaining in popularity.

2.2 HPLC

Despite the range of HPLC supports that are available (see pages 59–61), HPLC has not generally been used for the separation of free amino acids. This is partly because the normal cation exchange chromatography described above is so reliable that it has not been worthwhile to try to replace it by the more expensive HPLC apparatus; moreover, the rather complex ninhydrin detection system would still be needed, if the sensitivity were to be retained.

However, the separation of amino acids by reversed phase HPLC is particularly efficient, provided that any free amino groups are derivatised so that they are no longer basic. Furthermore, if the derivative is chosen to have a strong chromophore in the UV region, then rapid analysis by HPLC systems can be fast, efficient, and sensitive.

For example, the fluorenylmethoxycarbonyl (Fmoc) derivatives can be readily formed, and they absorb light strongly at about 280 nm.

Fluorenylmethoxycarbonyl chloride

(Fmoc—Cl)

Fmoc amino acid derivative

There are two advantages of carrying out amino acid analysis of suitable derivatives by HPLC:

(i) Standard HPLC equipment, including UV detection, can be used.
(ii) All the amino acids, including **proline**, can be detected readily.

2.3 Gas–Liquid Chromatography

In gas–liquid chromatography (GLC), an inert support (powdered glass or silica, or a fine glass capillary) is coated with an involatile organic layer (usually a hydrocarbon or a poly-alcohol), which will become liquid at elevated temperatures—this is the stationary phase, which is placed in a long (invariably coiled) glass tube. An inert gas (usually nitrogen) is then flowed through the heated column, and the sample to be analysed is injected into the gas stream in a suitable solvent. Without the packing in the column, all the components would simply be flushed through with the carrier gas. However, any affinity for the stationary liquid phase will cause those particular components to be eluted more slowly, and this allows the different compounds in a mixture to be separated.

Unfortunately, amino acids generally exist as zwitterions, and consequently they are relatively involatile. As only gaseous samples are carried through by the inert gas, free amino acids cannot be analysed by this technique. If the amino acids are derivatised, however, then they can be made relatively volatile, and can then be analysed by GLC.

Typical amino acid derivatives used for analysis by GLC. Either trifluoroacetamide esters (A) or silyl protected derivatives (B) are often used.—(NB. Any side-chains would be similarly derivatised.)

For amino acid analysis using GLC, the temperature of the column is usually increased with time; this is to ensure that the slowest-moving amino acid derivatives (usually the least volatile) still emerge reasonably quickly from the column.

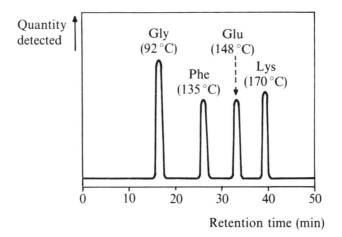

Retention time (min)

GLC analysis for the mixture of amino acids Glu, Gly, Lys, Phe. The trifluoracetamide esters were used [see (A) on page 75]. The column was run with a temperature gradient, and flame ionisation detection (f.i.d.) was used.

As the inert gas emerges from the GLC column, it can be readily analysed by flame ionisation detection (f.i.d.). For this, the eluting gas is fed into a flame, which is burning inside a chamber. A potential difference is applied across the chamber, and the current flowing will depend on the number of ionised particles produced by the flame.

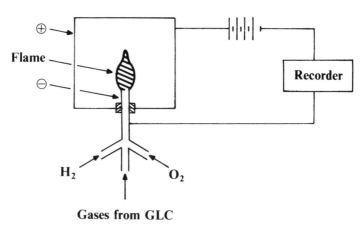

Figure 4.1. Flame ionisation detection (f.i.d.).

When an organic compound is eluted from the column, and burned in the gas flame, there is an increase in the number of ionised particles in the chamber; this causes an increase in the current, which is continuously monitored and recorded. This detection method is sensitive down to about 10^{-10} g of derivatised amino acid.

The advantages of GLC for amino acid analysis are as follows:

(i) Standard GLC equipment can be used.
(ii) **All** the amino acids can be readily detected; they will, however, give different f.i.d. responses, and this will need to be quantified.
(iii) GLC equipment can be linked directly to a mass spectrometer. This is particularly useful if unusual amino acids are being analysed, and can help in their identification.

2.4 Paper Electrophoresis

This technique simply involves loading the mixture of amino acids onto a strip of paper, which is then soaked in an aqueous buffer at a suitable pH (see pages 54–56). When a potential difference is applied to the paper, amino acids with a net positive charge migrate to the cathode (negative terminal), whilst those with a net negative charge migrate to the anode (positive terminal). For example, with a mixture of glycine, phenylalanine, glutamic acid, and lysine, we would expect to see the following electrophoretogram after running at pH 8, and spraying with ninhydrin solution.

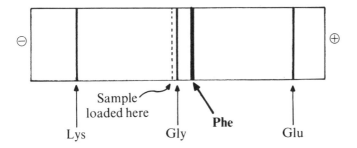

The separation of a mixture of four amino acids (Glu, Gly, Lys, Phe) using paper electrophoresis at pH 8.

It is easy to work out the results that you would expect at different pHs, simply by calculating the net charge (e.g. see page 38). This factor dominates the migratory aptitude of the amino acids, although the migration also depends on the size of the amino acid (plus its solvent shell): the bigger it is, the slower it will migrate.

The resolution from paper electrophoresis is much poorer than that from the other chromatographic techniques described above. Moreover, it is very difficult to carry out **quantitative** analysis on electrophoretograms. Nevertheless, there are some occasions when the technique is very valuable:

(i) It is sometimes used for the separation of unusual amino acids. If only a thin strip of the paper is developed with ninhydrin, then the position of the amino acids can be determined (see pages 63–64). After cutting off the appropriate bands, the amino acids can be washed from the paper and their structure determined by other methods.

(ii) Sometimes specific amino acids are radio-labelled (e.g. see pages 155–156). In these cases, the amino acids can be separated using paper electrophoresis, and their positions determined using ninhydrin; the radioactive amino acids can then be identified using a Geiger counter or scintillation counter.

2.5 Paper Chromatography

Paper chromatography is usually first encountered by chemists while still at school, for separating the dyes in black ink. Its use for the separation of amino acids relies on the fact that the cellulose groups in paper have a strong affinity for water molecules, which are held by hydrogen bonds to the OH groups of the polysaccharide chains. Therefore, if paper is used as the stationary phase, and an aqueous eluant is employed, the sample is effectively partitioned between two hydrophilic phases; the migration (or R_F) of the components will depend on their relative affinity for the two phases; the exact nature of the mobile phase is crucial, and typical eluants might contain:

water
ammonia solution (to increase pH)
ethanoic acid (to decrease pH)
butan-1-ol (to decrease dielectric constant)

If the amino acids cannot be fully separated by simple paper chromatography, then two-dimensional chromatograms can be tried. For this, the sample is applied as a spot, and the chromatogram is run in one direction with one eluant. The paper is then dried, the eluant is changed, and the chromatogram is run at right angles to the previous direction.

Using this tactic, it is relatively easy to separate the 20 DNA encoded amino acids; in the example below (Figure 4.2), a mixture of 10 of them have been readily resolved. As you can see, some of them would fail to separate if a single eluant was used.

Paper chromatography and paper electrophoresis are used in very similar situations; e.g. for separating unusual amino acids, or for identifying radio-labelled residues. Indeed, it is possible to run a two-dimensional analysis in which paper chromatography is performed in one direction, and electrophoresis is carried out at right angles.

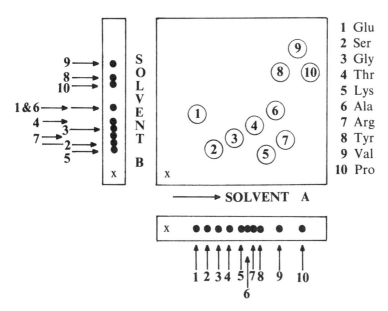

Figure 4.2. Two-dimensional paper chromatography of a mixture of 10 amino acids.

With all these permutations, it is usually possible to separate mixtures of amino acids using a suitable combination of electrophoresis at various pHs, and/or paper chromatography with a range of eluants.

Amino acid analysis does not usually allow the **absolute stereochemistry** of the amino acids to be determined (although DNA encoded ones are always L at the α-carbon). However, it is often possible to assign the configuration by:

(i) careful use of enzymic digestions;
(ii) chromatography of suitably derivatised amino acids;
(iii) comparison of the optical rotations with those of known amino acids.

---

3 Summary

Let us now return to our sample of PENTIN. At the end of Chapter 3, we had isolated 10 mg in a really pure state. We now need to determine which amino acid residues are present in it.

Therefore, we could take 100 μg of PENTIN, and subject it to total acidic hydrolysis. This would liberate the constituent amino acids. (We could check that tryptophan was **absent** by taking the UV spectrum of PENTIN.)

We could now carry out a standard amino acid analysis on our mixture of amino acids, using ion exchange chromatography ($ArSO_3^-$ support). The eluting amino acids could be detected and quantified by reaction with ninhydrin, and measurement of the absorption at 570 nm. We will assume that amino acid analysis gave the trace shown on page 73; comparison with standard amino acids would allow us to confirm that there were no unusual amino acids present in PENTIN, and that it was composed only of Glu, Gly, Lys, and Phe. In order to determine the **amounts** of each amino acid, we need to determine the area of each peak (integration), and compare this with the values for the standards; this would give the following results:

Amino acid	μmoles detected	Relative quantities (μmoles/0.216)*	Amino acid composition
Glu	0.213	0.99	1
Gly	0.441	2.04	2
Lys	0.207	0.96	1
Phe	0.219	1.01	1

Amino acid analysis of PENTIN after total acidic hydrolysis. The relative quantities (*) are found by dividing by *any* suitable number—0.216 gave values close to whole numbers, from which the amino acid composition could be readily seen.

> The total acidic hydrolysis of PENTIN would have cleaved all amide bonds. Glutamic acid would therefore have been liberated from either Glu or Gln residues—additional analysis will be needed in order to ascertain which is present in PENTIN.

The amino acid analysis reveals only the **relative** amounts of the amino acids present; the actual amino acid composition of PENTIN is therefore [Gly(2), Phe, Lys, Glu]$_n$, where n could be any whole number. But during the purification of PENTIN, it became apparent that the molecular weight was about 500. The molecular weight of a pentapeptide containing Gly(2), Phe, Lys, Glu would be 536. It is clear, therefore, that $n = 1$ for PENTIN, and its amino acid composition is:

[Gly(2), Phe, Lys, Glu]

So now we know which four amino acids are present in PENTIN. We also know the amounts of each amino acid that are present in one molecule of PENTIN. But we still don't know the **order** of the residues (there are 60 possibilities). In Chapter 5, we will see how we can determine the **sequence** of residues.

Further Reading

Laboratory Techniques in Biochemistry and Molecular Biology, Volume 9, *Sequencing of Peptides and Proteins*, 2nd Edition, G. Allen, Chapter 2, pp. 35–55, Elsevier, 1989. Strong practical bias.
Protein Sequence Determination, 2nd Edition, (S. B. Needleman, Ed.), Chapter 7, pp. 204–31, Springer-Verlag, 1975. A useful chapter.
Chemistry and Biochemistry of the Amino Acids, (G. C. Barrett, Ed.), Chapters 14/15/16, pp. 415–79, Chapman and Hall, 1985. Covers the separation of amino acids by a number of chromatographic methods.

Questions

1. After complete acidic hydrolysis, the hormone α-MSH gave the following amino acid trace after ion exchange chromatography:

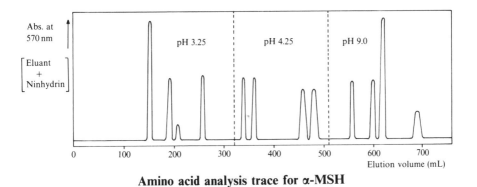

Amino acid analysis trace for α-MSH

(a) By comparison with the standards on page 72, deduce the amino acid composition of α-MSH.
(b) Given that α-MSH is DNA encoded, are there any details of its residues that are still unclear from its amino acid analysis?
(c) A strong UV absorption for α-MSH is observed at 280 nm; which amino acid (not detected above) might account for this, and how might its presence be confirmed?

2. A peptide is found to produce the following amino acids, after acidic hydrolysis and separation by electrophoresis:

(A)   (C)   (E)

(B)   (D)   (F)

Identification of the amino acids involved a combination of physical, chemical, and spectroscopic tests. One piece of information for each amino acid is given below; match each amino acid with the appropriate data.
(a) Is optically inactive
(b) Migrates to the anode*
(c) Co-migrates with a DNA encoded amino acid*
(d) Reacts with ninhydrin to give phenylethanone
(e) Is ninhydrin negative (gives yellow colour)
(f) Gives UV active adduct with sodium 4-chloromercuribenzoate
(* Electrophoresis at pH 6)

3. A peptide was isolated in ng quantities from leaf mold. The purification steps involved gel filtration (which indicated a MW of about 400) and electrophoresis at pH 7 (from which it was apparent that the peptide was uncharged). Although non-DNA encoded amino acids were likely to be present, amino acid analysis (after acidic hydrolysis) indicated the following residues:

Glu, Gly(2), Phe

Prolonged treatment of the peptide with aminopeptidase liberated only glycine. Similarly, reaction with carboxypeptidase over several hours liberated only a single amino acid, which was different from any of those isolated after acidic hydrolysis. Propose a structure for the peptide.

CHAPTER 5

Sequencing

Having determined the amino acids that are present in a peptide such as PENTIN, the real problem of structure elucidation starts—determining the **sequence** of the amino acids. This might not seem too difficult a task at first, but the snag is that every single amino acid is attached to the next one by a simple amide bond; so any conditions that might be intended to cleave one **specific** peptide bond could also result in the breakage of other peptide bonds too.

So how could we go about determining the order of the amino acids? To help us to think about this problem, we will assume that we have isolated another extremely important pentapeptide growth hormone, whose sequence eventually turns out to be:

Asp—Met—Phe—Lys—Ala

In case you cannot remember the side-chains for the abbreviated amino acids above, the full structure is drawn below.

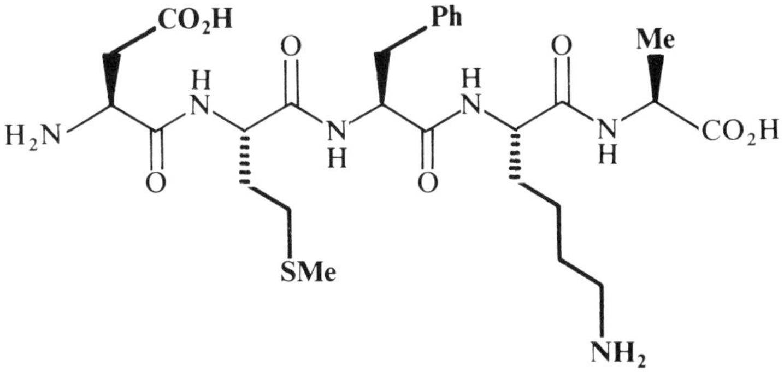

Asp—Met—Phe—Lys—Ala

We will consider the structure determination of this 'unknown' peptide in four sections.

1. Determination of just the amino- or carboxy-terminal residues.
2. Methods of sequencing peptides by chopping off one residue at a time.
3. Methods of breaking up a peptide into smaller fragments.
4. Other ways of solving the sequencing problem.

Finally, we will see how we might determine the amino acid sequence of PENTIN, using the methods developed in this chapter.

1 End Group Analysis

If we consider the amino acids that make up a peptide, it is clear that almost all of the amino and carboxylic acid groups will be actually involved in peptide bonds. The only exceptions are the C- and N-terminal groups, and possibly some of the side-chains (e.g. Lys, Asp, Glu). The simple tactic for identifying these terminal residues is to chemically modify them in some way that leaves the peptide bonds intact, and then to completely hydrolyse the peptide; the terminal amino acid would then be liberated in a modified form.

1.1 Amino Termini

Primary and secondary amino groups are usually quite nucleophilic, and this fact can be very easily exploited for end group analysis. For example, they react readily with the electrophilic 2,4-dinitrofluorobenzene (known as Sanger's reagent, after the scientist who developed its use), to give 2,4-dinitrophenyl (or DNP) derivatives.

2,4-Dinitrofluorobenzene
(Sanger's reagent)

Anionic intermediate stabilised by delocalisation. e.g.

If this was carried out for the 'unknown' pentapeptide at the beginning of this chapter, then the N-terminal Asp would be present as the DNP derivative.

Having formed this DNP derivative, the peptide can be completely hydrolysed using concentrated HCl (see pages 67–69). The DNP group is stable to acid, and so the amino terminal residue would be liberated as its DNP derivative; all other residues would be present as the free amino acids.

For DNA encoded amino acids, the DNP derivative can be readily identified by comparison with standard derivatives using chromatography (e.g. HPLC retention times; see pages 59–61). This is particularly straightforward because the DNP derivatives absorb strongly at 360 nm (and consequently are a yellow colour). In the case of our 'unknown' peptide, the N-terminal amino acid is aspartic acid (Asp), and this would be identified as its DNP derivative. One other residue would have also formed a DNP derivative; because of its amino side-chain, lysine (Lys) would also have reacted with Sanger's reagent, but this side-chain derivative could also have been readily identified by comparison with standards.

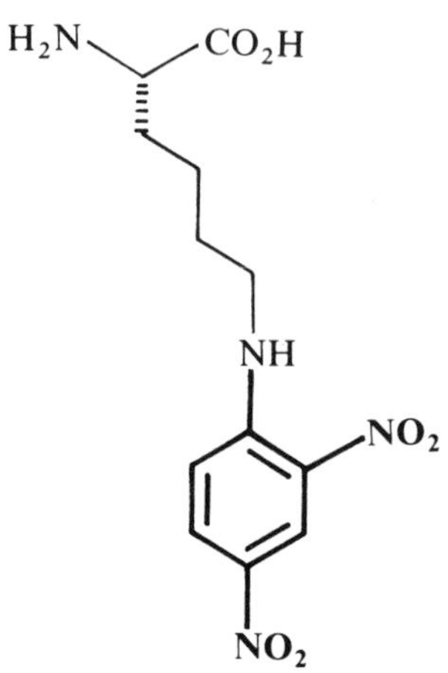

Lys (DNP)

DNP—Lys (DNP)

The N$^\varepsilon$-dinitrophenyl derivative of lysine is generated from *any* Lys residues in peptides (except the N-terminus).

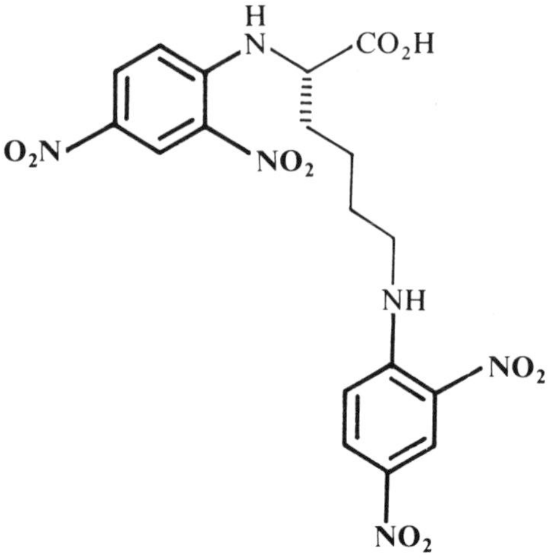

If lysine is the N-terminal residue of a peptide, the N$^\alpha$,N$^\varepsilon$-*bis*-DNP derivative is formed.

The identification process can be improved by a simple extraction procedure. Under mildly acidic conditions, the N-terminal DNP derivative would be soluble in simple organic solvents such as trichloromethane or ethyl ethanoate (its nitrogen being relatively non-basic due to conjugation), whereas all of the other liberated amino acids will possess a basic amino group, and will therefore remain in the aqueous phase. Therefore, the N-terminal DNP derivative can be readily isolated in a fairly pure state [even Lys(DNP) would be removed], making identification particularly easy.

More recently, dansyl chloride has largely superseded Sanger's reagent. Thus 5-(dimethylamino)naphthylsulphonyl chloride also reacts with primary and secondary amino groups, giving sulphonamides that are acid stable.

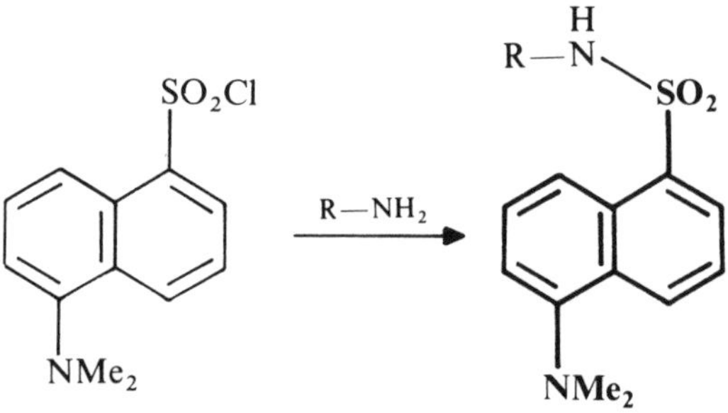

Dansyl chloride

Reaction of a peptide with dansyl chloride, hydrolysis of the peptide, and isolation of the dansyl derivatised amino acid can be carried out exactly as for Sanger's reagent. However, the dansyl derivatives have the additional advantage that they absorb very strongly in the UV region, and it is therefore possible to detect minute quantities of dansyl derivatives; this means that less than $1\,\mu g$ ($10^{-8}\,mol$) of an amino acid can be easily detected and identified.

1.2 Carboxy Termini

The identification of the carboxy terminus is generally less easy than the N-terminus. There are two approaches that can be employed.

(i) Treat the peptide with hydrazine. Hydrazine is a powerful nucleophile, and will cleave carboxylic acid derivatives (e.g. amide bonds) to give the corresponding acid hydrazides. But hydrazine is also a strong base; this means that any free carboxylic acid groups are converted into the corresponding carboxylates, and are no longer susceptible to nucleophilic attack. Thus, if we consider alanine as the C-terminal amino acid, then the following bond cleavage occurs:

Eventually, after prolonged heating with hydrazine, all of the constituent amino acids in the peptide are converted into their acid hydrazides except for the C-terminal residue, which will be present as the carboxylate. This can be extracted into dilute aqueous alkali, and then identified by amino acid analysis (see pages 70–79).

(ii) Reduction of the ester with $LiBH_4$. If any free carboxylic acid groups are converted into their corresponding methyl esters, then it is possible to reduce the ester groups without attacking the peptide bonds.

> Lithium aluminium hydride reduces both esters and amides down to the lowest oxidation levels (alcohols and amines respectively).
>
> Sodium borohydride will not readily reduce esters or amides.
>
> Lithium borohydride is just a little more reactive than $NaBH_4$, and is able to selectively reduce esters in the presence of amides.

So, if the following peptide were esterified, and subsequently reacted with lithium borohydride, then the transformations shown below would take place:

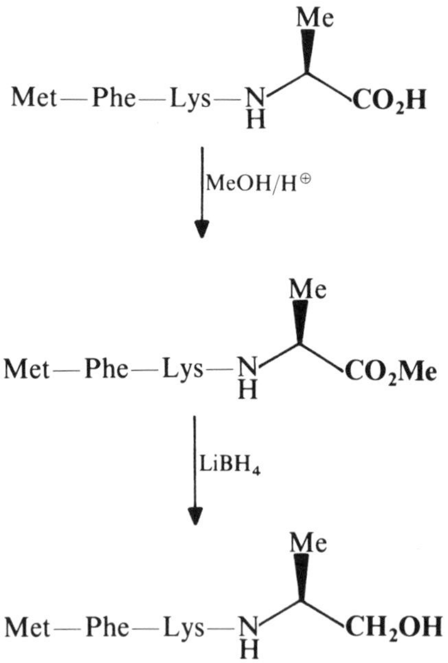

Me

Met—Phe—Lys—N—CH(Me)—CO_2H
            H

$MeOH/H^{\oplus}$

Me

Met—Phe—Lys—N—CH(Me)—CO_2Me
            H

$LiBH_4$

Me

Met—Phe—Lys—N—CH(Me)—CH_2OH
            H

If the modified peptide was then submitted to total acidic hydrolysis, then free amino acids would be liberated for all residues except the C-terminus—which would be generated as the amino alcohol:

$$\text{Met—Phe—Lys—N} \underset{\text{H}}{\overset{\text{Me}}{\diagdown}} \text{CH}_2\text{OH}$$

$$\downarrow \text{H}^\oplus/\text{H}_2\text{O}/\text{Heat}$$

$$\boxed{\text{H}_2\text{N} \overset{\text{Me}}{\diagup} \text{CH}_2\text{OH}}$$

+
Met
+
Phe
+
Lys

The solitary amino alcohol (reduced alanine in this case) could be extracted into an organic phase (e.g. ethoxyethane) at pH 11, whereas the amino acids would remain in the aqueous phase. Identification of the amino alcohol would thereby assign the C-terminal residue.

In fact, neither of the above methods is particularly sensitive, with a lower limit of about 10 μg. Because of this, special radio-labelling methods of determining the C-terminus have been developed (see pages 155–156).

2 Chopping Off One Residue at a Time

Perhaps surprisingly, it is usually possible to remove just one residue at a time, using either chemical or enzymic methods. This is, of course, the ideal way of determining the sequence of a peptide.

2.1 Edman Degradation

This is the most common method of sequencing peptides, and starts from the free N-terminus. The reagent used is phenylisothiocyanate (Ph—N=C=S); the mechanism may at first look a bit complicated, but it actually consists of two fairly easy steps.

Firstly, the free amino group reacts with the Edman reagent, just as if the C=N were a carbonyl group, giving a thiourea derivative.

Secondly, 6M hydrochloric acid is added to the reaction mixture. Under these conditions, any excess Ph—N=C=S is rapidly destroyed by hydrolysis. The acid then promotes the formation of a thiohydantoin, as if the nitrogen (marked *) had attacked the peptide bond.

Note: In actual fact, the more nucleophilic sulphur atom **initially** attacks the adjacent peptide bond, and gives the five-membered sulphur-containing ring; under the reaction conditions, this is converted into the **thermodynamically preferred** thiohydantoin. If you're feeling keen, you might like to work out how this interconversion occurs, although these details don't affect the result of the reaction.

> The acidification step is the crucial one, which leads to selective cleavage of the N-terminal residue. As with many selective reactions in peptide chemistry, it relies on the fact that five-membered rings are formed extremely rapidly. This step only occurs after acidification, and so is presumably triggered by protonation of the amide nitrogen or oxygen.

The final products from the reaction are a **thiohydantoin** derivative, and a new shortened peptide **with a free N-terminus**. If, for example, we consider our pentapeptide Asp—Met—Phe—Lys—Ala, then the thiohydantoin derivative would be as follows:

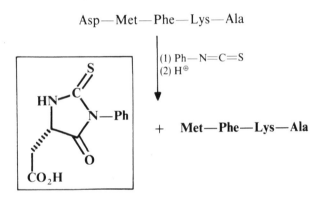

We would also have formed a new tetrapeptide, in which methionine would be the N-terminus:

Whilst the new peptide should be readily soluble in water, we could extract the thiohydantoin into an organic phase, and identify it by comparison with known thiohydantoins. Just as importantly, we could isolate the water-soluble tetrapeptide, and repeat the Edman degradation. This would reveal methionine as the next amino acid, and would generate a new tripeptide. Indeed, even if we possessed only 50 μg of our original pentapeptide, we would be able to sequence it entirely using Edman's method. And the fact that the sequence of reactions is repeated exactly for each residue means that it is now possible to automate the whole procedure.

The Edman degradation is undoubtedly the most important method of sequencing peptides, but there are a few limitations. Firstly, it is not as sensitive as is sometimes required, when only minute amounts of a peptide are available; this is why end group analysis is still of value—and an example of combining Edman and N-terminal analysis is discussed in Chapter 7. Secondly, the Edman degradation requires two chemical steps for cleavage of the N-terminal residue; the resulting shortened peptides slowly accumulate impurities as the cycle is repeated, and unambiguous analysis is only viable for 20 or so residues. For bigger polypeptides, the molecule must first be broken down into manageable chunks (see pages 93–99), or sequencing of the genetic code may be more appropriate (see pages 211–219). Finally, the Edman degradation requires that the N-terminus contains a **free amino group**; many naturally occurring peptides do not meet this requirement, and other tactics must be employed (see later in this chapter).

2.2 Carboxypeptidase

There are two types of enzyme that can be used to completely hydrolyse peptides into their constituent L-amino acids. Aminopeptidase was mentioned in Chapter 4 (pages 69–70), and it rapidly cleaves off amino acids sequentially, starting at the amino terminus. Carboxypeptidase sequentially removes one amino acid at a time, but starting from the C-terminus. But this hydrolysis is fairly slow, and so it is often possible to determine the first few C-terminal residues by quenching the enzymic hydrolysis after varying times (e.g. by acidification), and then analysing for the amino acids liberated.

Peptidases are enzymes that cause the sequential cleavage of peptide bonds. Aminopeptidase starts from the amino terminus, and carboxypeptidase from the carboxy terminus. Hydrolysis does not usually proceed past proline residues or D-amino acids.

The rate of hydrolysis varies considerably between different peptide bonds, so the procedure requires some trial and error. But a typical experiment might yield the following results with our Asp—Met—Phe—Lys—Ala pentapeptide.

Reaction time (min.)	Liberated amino acids (mol. equiv.)
2	Ala (0.34)
6	Ala (0.73), Lys (0.17)
20	Ala (0.96), Lys (0.52), Phe (0.07)
60	Ala (0.98), Lys (0.72), Phe (0.18), Met (0.12)

As the reaction time is increased, so the variations in the rates of hydrolysis create a more and more complicated picture. Thus, although this can be a very informative method, particularly if the N-terminus is blocked (and so unavailable for Edman degradation), it is normally only possible to sequence two or three of the carboxy terminal residues.

3   Methods of Dissecting Peptides

There are two possible reasons for wishing to cleave a peptide into smaller fragments:

(i) There might be no available N- or C-termini from which to start a sequencing experiment. This is a surprisingly common problem; it might occur if the peptide were cyclic, or if both the N- and C-termini were blocked, as exemplified below:

$$Ac—Asp—Met—Phe—Lys—Ala—NH_2$$

But if we could cleave the Met—Phe peptide bond, for example, then we would obtain a dipeptide with a free C-terminus, and a tripeptide with a free N-terminus:

Ac—Asp—Met

Phe—Lys—Ala—NH$_2$

We could then use the methods described above in order to sequence these fragments.

(ii) Suppose that we possessed a polypeptide of more than 20 residues. However carefully we conducted the Edman degradation, it is unlikely that we would be able to sequence the entire peptide in one go. But if we could cleave it into fragments of less than 20 residues, then we could sequence each of the fragments separately; provided that we could piece the sections back together in the correct order, we would be able to deduce the sequence of the original polypeptide.

Cleavage of a specific peptide bond might be achieved using either chemical or enzymic methods. Of course, enzymes are merely very specific chemical catalysts, with their selectivity being induced by the three-dimensional shape of the active site. In the chemical cleavage of peptide bonds, on the other hand, an adjacent side-chain is actually involved in the chemistry that leads to fragmentation of the peptide bond.

3.1 Enzymic Cleavage

There are an enormous range of enzymes that can be used for the cleavage of specific peptide bonds. This is partly because animals need peptidases in order to digest protein; but, more specifically, many organisms use peptidases to control hormonal responses.

For example, our body naturally produces a decapeptide called angiotensin I, which is biologically inert; if our body requires more blood to circulate, then it can produce an enzyme called angiotensin converting enzyme (ACE), which specifically cleaves an Phe–His bond in angiotensin I. The resulting octapeptide is called angiotensin II, and it causes a rapid increase in blood pressure.

Asp—Arg—Val—Tyr—Ile—His—Pro—Phe⦃His—Leu

Angiotensin I   **ACE**

Angiotensin Converting Enzyme (ACE)

Asp—Arg—Val—Tyr—Ile—His—Pro—Phe (+ His—Leu)

Angiotensin II

Causes hypertension (high blood pressure)

Small amounts of ACE can process large quantities of angiotensin II, so the body is able to respond quite rapidly to demand for more blood. Other peptidases degrade angiotensin II, and are thereby involved in controlling blood pressure.

In fact, there are hundreds of peptidases which cleave different peptide bonds, with varying degrees of specificity. However, in the determination of peptide sequences, there are a small number of enzymes that are particularly widely used, and their points of cleavage are worth remembering:

Enzyme	Site of cleavage	Residues
Trypsin	Carboxy side of basic residues	Arg, Lys
Chymotrypsin	Carboxy side of aromatic residues	Phe, Tyr, Trp
Thermolysin	Amino side of hydrophobic residues	Phe, Trp, Leu

The specificity of these enzymes is due to the shape of the active site, as exemplified in Figure 5.1 on page 96.

There is no guarantee that the enzymes will cleave the expected bond in all peptides—sometimes the structure of the peptide prevents this happening—in which case other peptidases can be tried. But trypsin, chymotrypsin, and thermolysin are commonly employed because they are fairly reliable.

So, if we consider the pentapeptide

$$\text{Ac—Asp—Met—Phe—Lys—Ala—NH}_2$$

in which sequencing is prevented because the N- and C-termini are both blocked,

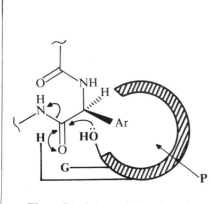

Figure 5.1. Schematic drawing of chymotrypsin.

Specific peptide bond cleavage by chymotrypsin relies on several features of the enzyme:

(i) Hydrophobic pocket (**P**) to bind specific side-chains.

(ii) Hydrogen bond to the carbonyl group (**G**), weakening the C=O π-bond.

(iii) A histidine residue (**H**) that readily transfers a proton to the amine leaving group.

(iv) The **OH** on a serine residue, which actually cleaves the amide bond; subsequent hydrolysis with water liberates the two peptide fragments, and regenerates the free enzyme, ready to catalyse the hydrolysis of another peptide bond.

we could use any of the enzymes above to create two fragments; these could then be sequenced, in order to determine the primary structure of the original pentapeptide.

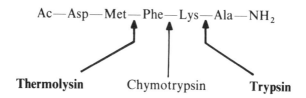

3.2 Chemical Cleavage

This relies almost invariably upon an ability to use a nearby side-chain in the formation of a five-membered ring. You should remember that such reactions are often kinetically favoured, and the two examples chosen below use this principle.

3.2.1 *Mild Acid Hydrolysis.* Typical conditions are 6M hydrochloric acid at room temperature for three days—not very subtle, but most peptide bonds are remarkably stable to this treatment. However, those residues that contain a nucleophilic side-chain can employ neighbouring group participation in peptide bond cleavage, and the peptide bond on the carboxy side of aspartic acid residues are particularly susceptible.

As you can see, reaction is believed to proceed via the cyclic anhydride. Glutamic acid also increases the rate of hydrolysis of adjacent peptide bonds (via the six-membered anhydride), although to a lesser extent.

Our fully blocked pentapeptide would be cleaved as follows:

3.2.2 *Cyanogen Bromide.* This reagent causes peptide bond cleavage on the carboxy side of methionine residues, again via a five-membered ring. The first step involves displacement of bromide by the very nucleophilic sulphur atom.

The carbon next to the sulphur has now become susceptible to nucleophilic attack, and the carbonyl oxygen of the amide (see page 5) is able to act as a nucleophile.

The product now possesses an imine function, which is readily hydrolysed to amine plus carbonyl group upon the addition of water:

The overall reaction generates a δ-lactone (i.e. five-membered cyclic ester) in place of the original methionine; the hydrolysed peptide bond possesses a free N-terminus, and so is amenable to end group analysis or sequencing. For example, our blocked pentapeptide would generate the following fragments:

$$Ac—Asp—Met—Phe—Lys—Ala—NH_2$$

(1) BrCN
(2) H_2O

Cyanogen bromide is quite widely used not only because it is a very reliable hydrolytic reagent, but also because the frequency with which Met is present in polypeptides and proteins often yields fragments of 20 or so residues—ideally suited for sequencing by Edman degradation.

(By using bromine, a similar reaction sequence can be used to cleave the peptide bond on the carboxy side of tryptophan or tyrosine; a bromonium intermediate is attacked by the amide oxygen to yield a five-membered ring, hydrolysis of which causes cleavage of the peptide bond involved.)

Disulphide bonds. Many peptides possess S—S bonds between cysteine residues (see page 14). These can be cleaved by the addition of mercaptoethanol ($HS-CH_2CH_2-OH$) or performic acid (HCO_2OH):

If a peptide is unaltered by the addition of mercaptoethanol, then it does not contain S—S bonds.

If several S—S bonds are possible in a peptide, then the cystine (Cys—Cys) can often be located as follows:

  (i) Oxidise all cysteine/cysteine with performic acid, and sequence the resulting peptide (e.g. using Edman degradation).

 (ii) Use an enzymic or chemical method to cleave the original peptide (with S—S bonds intact) at two or three sites, and separate the resulting peptide fragments.

(iii) Amino acid analysis of the fragments should reveal (with luck) those sections of the original peptide that must have been linked by an S—S bond.

e.g.   Ala—Cys—Phe—Cys—Phe—Cys—Glu

   (i) Oxidise (HCO_2OH) and sequence.
  (ii) Hydrolyse original peptide with chymotrypsin ($\uparrow$) $\rightarrow 2$ fragments; separate.
 (iii) Amino acid analysis of one fragment $\rightarrow$ [Ala, Cys_2, Glu, Phe]; Ala and Glu could only be in same pentapeptide if adjacent Cys residues were linked by a disulphide bond.

4 Other Sequencing Methods

There are many variations and extensions of the ideas outlined above for the sequencing of peptides, and some of these are mentioned in Chapter 7. And for DNA encoded peptides, nucleotide sequencing is also a possibility; see Appendix C. But two completely different methods that lead to the direct sequencing of peptides, and which are being used more and more frequently, are mass spectrometry and NMR spectroscopy.

4.1 Mass Spectrometry

The success of the Edman degradation procedure relies upon the sequential removal of amino acid residues from a peptide—but a lot of chemical steps are necessary. With mass spectrometry, the same result can be achieved (if all goes well) from a single spectrum.

Mass spectrometry initially requires that the sample be ionised. This is often achieved by bombarding it with particles (e.g. electrons); loss of an electron from the sample then generates a radical cation. The cation is then accelerated across a potential difference, and its mass determined by the extent to which its path is bent by a magnetic field. This mass might correspond to the molecular weight of the parent molecule, or fragmentation might occur, giving lower molecular weight components.

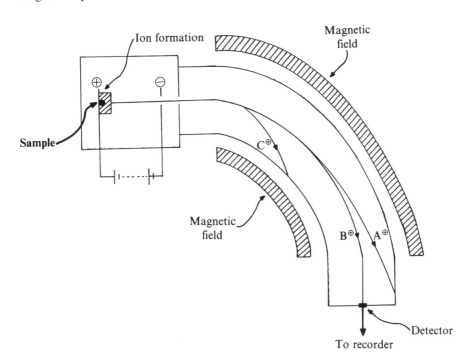

Figure 5.2. The principles of mass spectrometry.

For peptides, the amide oxygen is one of the most likely atoms to become ionised, and a relatively simple fragmentation can then occur, which results in cleavage of the adjacent peptide bond.

It is not hard to show that our pentapeptide might be expected to give a mass spectrum that would allow the entire sequence to be determined from one experiment. However, there are two major problems:

(i) Most peptides are not sufficiently volatile for them to desorb at an appreciable rate from the mass spectrometer sample probe, so that the spectra are too weak to give meaningful results. Derivatising the peptide, or raising the temperature of the probe, can increase the intensity of the spectrum.
(ii) Fragmentations other than those described above are usually possible, leading to extremely complex spectra. This is particularly true if additional energy is given to the molecule during formation of the radical cation—for example, if the sample is heated (see above) or if high-energy electrons are used to ionise the sample.

However, one superb method of ionisation has been developed which not only gives good sensitivity with relatively involatile substances, but which also transfers very little energy to the resulting ion, so that only the simplest fragmentations occur. This is called fast atom bombardment (FAB) mass spectrometry.

In FAB mass spectrometry, atoms of an elemental gas are bombarded with electrons, causing ionisation to radical cations. These ions are accelerated through a potential difference, and are then allowed to collide with more of the (neutral) gas in a second chamber, generating neutral (but fast-moving) atoms of the element. Any remaining charged species are removed by a transverse magnetic field, whilst the fast-moving neutral atoms are directed at the sample.

Fast atom 'gun' for FAB mass spectrometry

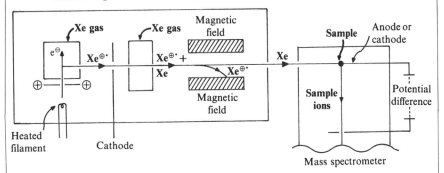

On collision, the fast-moving atoms are able to volatilise the sample molecules without transferring much excess energy to them. Any such molecules that have picked up a proton (from the sample matrix) will produce a mass spectrum, but the fragmentation pattern will be very simple:

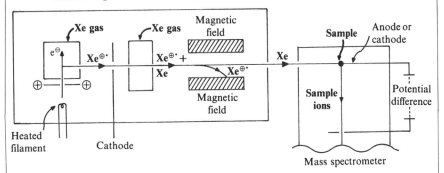

Peptide Bond Fragmentation after FAB. The same major fragmentations occur as from a radical cation (e.g. after ionisation by electron impact). However, the lower-energy cation from FAB undergoes fewer unexpected fragmentations, leading to a much cleaner spectrum.

This FAB technique has allowed the sequencing of several particularly awkward peptides. The expected FAB peaks for our pentapeptide are shown below:

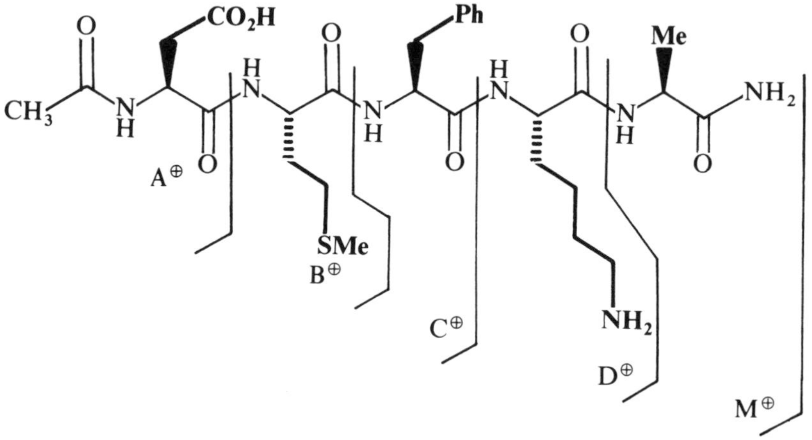

Fragment	Formula	m/z	− CO
A⊕	$C_6H_8NO_4$	**158**	130
B⊕	$C_{11}H_{17}N_2O_5S$	**289**	261
C⊕	$C_{20}H_{26}N_3O_6S$	**436**	408
D⊕	$C_{26}H_{38}N_5O_7S$	**564**	536
M⊕	$C_{29}H_{45}N_7O_8S$	**651**	

An important feature of FAB mass spectrometry is that peptides containing unusual amino acids can often be analysed; the unusual residues often give characteristic spectra, whereas standard amino acid analysis requires standard samples for comparison. However, even using FAB, peptides containing more than about 10 residues usually undergo too many unexpected fragmentations to allow an unambiguous determination of the sequence. Nevertheless, mass spectra often yield important information that aids sequencing (e.g. see Chapter 7), it requires very little material (a few micrograms should be enough), and sequencing of short peptides can sometimes be achieved in a single experiment using FAB mass spectrometry.

4.2 NMR Spectroscopy

Although NMR spectroscopy is one of the most powerful modern methods of structure determination, it is rarely used for the sequencing of peptides. There are a number of reasons for this:

(i) Relatively large quantities of peptides are required—usually several milligrams of pure peptide (cf. microgram quantities for most other methods).

(ii) It is often difficult to obtain really sharp spectra of peptides—solvent and temperature can be critical—and most peptides give quite complex NMR results.

(iii) Even with sharp spectra, sequencing a peptide from NMR results is still extremely difficult. This is because each residue is a self-contained unit (as far as simple ^{1}H or ^{13}C NMR spectra are concerned). The way round this problem is to use special NMR pulse sequences that either identify long-range couplings between residues, or that identify atoms that are close in space (using the nuclear Overhauser effect or NOE); the NOE technique gives important information about the three-dimensional structure of peptides, as well as about the amino acid sequence (see Appendix B).

5 Summary

Returning to our anti-arthritic peptide, PENTIN, you may recall that total acidic hydrolysis revealed that it was a pentapeptide composed of:

[Gly(2), Phe, Lys, Glu]

In order to sequence PENTIN, we might start with an end group analysis. After treatment of 10 μg of PENTIN with dansyl chloride followed by acidic hydrolysis, the only strongly UV absorbing product is the side-chain dansyl derivative of lysine; so the N-terminus of PENTIN must be blocked, and Edman degradation will not be possible.

PENTIN $\xrightarrow[\text{(2) H}^{\oplus}\text{/H}_2\text{O/Heat}]{\text{(1)}}$

(1) Dansyl chloride (with SO$_2$Cl and NMe$_2$ naphthalene structure)

$\longrightarrow$ H$_2$N—CO$_2$H + [Glu, Gly (2), Phe]

(structure with NH—SO$_2$—naphthalene—NMe$_2$)

   Treatment of 10 μg of PENTIN with carboxypeptidase leads to the liberation of the constituent amino acids—so we know that the C-terminus is present as the free carboxylic acid. As our amounts of PENTIN are relatively small, we might ignore end group analysis of the C-terminus for the time being. But PENTIN contains two residues that might allow us to fragment it enzymically:

> The **lysine** residue might direct cleavage by **trypsin**—when this is tried on 10 μg of PENTIN, the intact pentapeptide is recovered by HPLC. The **phenylalanine** residue might direct cleavage by **chymotrypsin**—this time, reaction of a 10 μg sample of PENTIN yields two components, readily separable by HPLC.

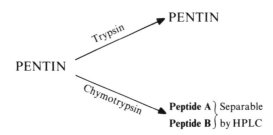

PENTIN
Trypsin → PENTIN
Chymotrypsin → Peptide A }
Peptide B } Separable by HPLC

   In order to have enough material to handle, it would be sensible to repeat the chymotrypsin reaction on (say) 50 μg. Using HPLC, the two fragments can be isolated, and separately submitted to Edman degradation. Only one of the peptides yields a thiohydantoin derivative, which can be identified as the Gly derivative by its gas chromatography (GC) retention time.

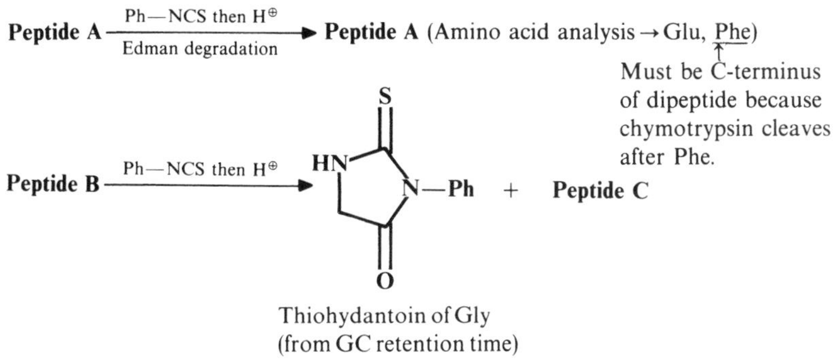

Peptide A $\xrightarrow[\text{Edman degradation}]{\text{Ph—NCS then H}^{\oplus}}$ Peptide A (Amino acid analysis → Glu, Phe)

Must be C-terminus of dipeptide because chymotrypsin cleaves after Phe.

Peptide B $\xrightarrow{\text{Ph—NCS then H}^{\oplus}}$ [thiohydantoin structure] + Peptide C

Thiohydantoin of Gly
(from GC retention time)

A further Edman degradation again yields the thiohydantoin derivative of glycine, with the other product identifiable as lysine (by its reaction with dansyl chloride to give the *bis*-dansyl derivative).

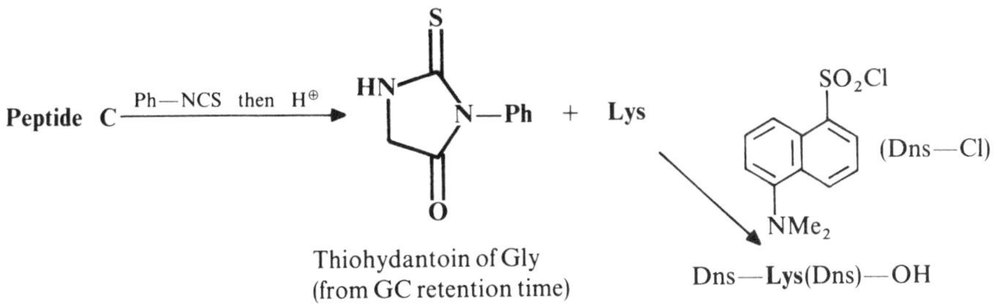

Peptide C $\xrightarrow{\text{Ph—NCS then H}^{\oplus}}$ [thiohydantoin structure] + Lys

(Dns—Cl)

Thiohydantoin of Gly
(from GC retention time)

Dns—Lys(Dns)—OH

If we now piece together all of the information, we can deduce the amino acid sequence of PENTIN:

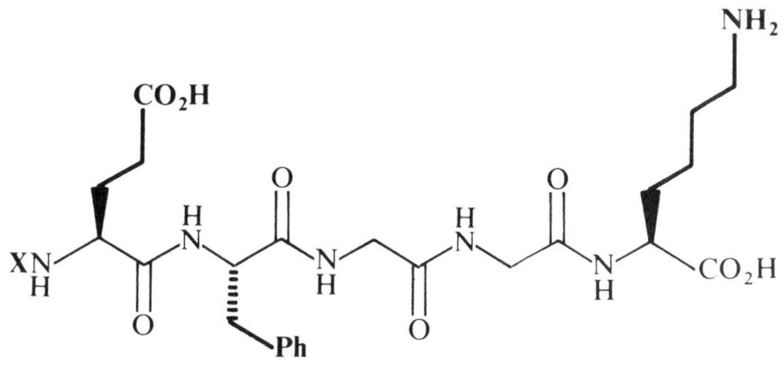

X—Glu—Phe—Gly—Gly—Lys

Finally, we need to know how the N-terminus is blocked. With so little PENTIN available, NMR does not seem particularly hopeful. But mass spectrometry only requires a few micrograms. With FAB mass spectrometry, the higher molecular weight regions are rather complex for PENTIN, but the lower molecular weight peaks (which might be expected to yield information about the N-terminus) show major signals at m/z 112 and 84.

$$\text{PENTIN} \xrightarrow[\text{spectrometry}]{\text{FAB mass}} \text{Peaks for } m/z \text{ 112/84}$$

Unit	Formula	MW
Glu	$C_5H_9NO_4$	147
Glu—	$C_5H_8NO_3$	130
Glu(– H_2O)—	$C_5H_6NO_2$	**112**

Only the cyclic amide derivative of glutamic acid (abbreviated to Glp, see page 29) gives these characteristic peaks, and must therefore be the N-terminal residue of PENTIN.

Despite all the operations required for the sequencing of PENTIN, less than $100 \mu g$ would be required in order to elucidate the following primary structure:

Glp—Phe—Gly—Gly—Lys

Finally, we need to synthesise PENTIN ourselves. This will allow us to confirm that we have determined its structure correctly—and will also enable us to make larger quantities for the testing of its anti-arthritic properties.

Further Reading

Protein Sequence Determination, 2nd Edition, (S. B. Needleman, Ed.), Springer-Verlag, 1975. A very useful text, with three particularly relevant sections:
Chapter 3 (pp. 30–103), End group analysis
Chapter 5 (pp. 114–61), Fragmentation of peptides
Chapter 8 (pp. 232–79), Sequence determination
Laboratory Techniques in Biochemistry and Molecular Biology, Volume 9, *Sequencing of Peptides and Proteins*, 2nd Edition, G. Allen, Elsevier, 1989. The practical side of sequencing is covered in three main sections:
Chapter 2 (pp. 61–71), End group analysis
Chapter 3 (pp. 73–104), Fragmentation of peptides
Chapter 6 (pp. 207–293), Sequence determination
The Proteins; Composition, Structure, and Function, Volume 3, 3rd Edition (H. Neurath and R. L. Hill, Eds.), Academic Press, 1978. Part of a very thorough coverage of protein chemistry and biochemistry.

Questions

1. Determine the structure of the following DNA encoded tripeptides, based on their amino acid composition and end group analysis.

(a) **A** [Gly, His, Phe]

$$\textbf{A} \xrightarrow[\text{(2) H}^+/\text{H}_2\text{O/Heat}]{\text{(1) 2,4-Dinitrofluorobenzene}} \text{DNP—His}$$

$$\textbf{A} \xrightarrow{\text{H}_2\text{N—NH}_2} \text{Gly (as the only free amino acid)}$$

(b) **B** [Ala, Asp, Pro]
Treatment with carboxypeptidase yields only Asn.
B is ninhydrin positive.

(c) **C** [Lys, Gly(2)]

$$\textbf{C} \xrightarrow[\text{(2) H}^+/\text{H}_2\text{O/Heat}]{\text{(1) Dansyl chloride}} \overset{\overset{\textstyle\text{Dns}}{\textstyle|}}{\text{Dns—Lys—OH}}$$

Carboxypeptidase fails to liberate any amino acids.

(d) **D** [Ala, Glu, Gly]

$$\textbf{D} \xrightarrow[\text{(3) H}^+/\text{H}_2\text{O/Heat (4) Basify and extract into EtOAc}]{\text{(1) CH}_2\text{N}_2 \text{ (2) LiBH}_4} \quad \text{(structures)}$$

$$\textbf{D} \xrightarrow[\text{(2) H}^+/\text{H}_2\text{O/Heat}]{\text{(1) HNO}_2\text{(aq)}} \text{Ala + Glu (as the only ninhydrin positive products)}$$

2. A hypertensive decapeptide **E** was obtained from ox serum. Deduce its structure from the information given below:

(a) It was composed of the following amino acids: [Arg, Asp, His(2), Leu, Phe, Pro, Tyr, Val(2)]

(b) Treatment of **E** with Sanger's reagent (2,4-dinitrofluorobenzene), followed by total acidic hydrolysis, gave DNP—Asp.

(c) Treatment of **E** with carboxypeptidase gave Leu, together with traces of Phe and His.

(d) Treatment of **E** with trypsin gave a dipeptide **F** and an octapeptide **G**, which were separated by paper chromatography. Amino acid analysis of **F** identified [Arg, Asp], whilst treatment of **G** with Sanger's reagent, followed by total acidic hydrolysis, gave the DNP derivative of Val.

(e) Digestion with chymotrypsin gave three fragments (**H**, **J**, and **K**), which were separated by paper electrophoresis. Amino acid analysis of them gave the following results:

H [Arg, Asp, Tyr, Val]

J [His, Phe, Pro, Val]

K [His, Leu]

(f) Reaction of **J** with Edman's reagent (PhNCS) liberated, successively, the phenylthiohydantoin (PTH) derivatives of Val, His, and Pro.

3. Neurotensin (**K**) is an important peptide involved in the transmission of nerve impulses. Deduce the amino acid sequence from the results below.

(a) After acidic hydrolysis, amino acid analysis revealed:

[Arg(2), Asp, Glu(2), Ile, Leu(2), Lys, Pro(2), Tyr(2)]

(b) Reaction of **K** with dansyl chloride, followed by acidic hydrolysis, gave only the N^ε-dansyl derivative of lysine, and no N^α derivatives.

(c) Mass spectrometry gave (among other peaks) a major fragment with m/z 112.

(d) Enzymic hydrolysis of **K** with carboxypeptidase was quenched after varying times, with the liberation of amino acids in the following amounts:

Time	Amino acids
2 min.	Ile(0.32), Leu(0.46)
6 min.	Ile(0.50), Leu(0.65)
16 hours	Ile(1.02), Leu(0.94), Tyr(0.95)

(e) Enzymic hydrolysis of **K** with chymotrypsin gave three peptides (**L**, **M**, and **N**), which had the following amino acid analyses:

L [Glu, Leu, Tyr]

M [Arg(2), Asp, Glu, Lys, Pro(2) Tyr]

N [Ile, Leu]

(f) Enzymic hydrolysis of **K** with papain gave three peptides (**P**, **Q** and **R**), which had the following amino acid analyses:

P [Glu(2), Leu, Tyr]

Q [Arg, Asp, Lys, Pro]

R [Arg, Ile, Leu, Pro, Tyr]

(g) Edman degradation of **Q** gave the thiohydantoin derivatives of Asp, then Lys; treatment of **Q** with carboxypeptidase liberated Arg.

(h) Edman degradation of **R** gave the thiohydantoin derivatives of Arg, then Pro.

4. The total acidic hydrolysis of a peptide (**S**) gives the following L-amino acids in equal amounts:

[Asp, Cys, Leu, Lys, Met, Val, Trp]

Mild acid hydrolysis of **S** with dilute HCl gives a heptapeptide **T** as the major product. Treatment of **T** with Sanger's reagent (2,4-dinitrofluorobenzene), followed by total acidic hydrolysis, gives DNP—Leu and N^ε-DNP—Lys in equal amounts.

Hydrolysis of **S** catalysed by chymotrypsin gives a heptapeptide **U** which, after treatment with Sanger's reagent and acidic work-up, gives di-DNP—Lys. Mild acid hydrolysis of **U** with dilute HCl gives mainly a tripeptide **V** and a tetrapeptide **W**. **V** exhibits an intense UV absorption (280 nm), and reacts with 1 mole of mercuribenzoate.

Hydrolysis of **S** catalysed by trypsin gives a heptapeptide **X** which, after the

Sanger procedure, gives DNP—Val and N^ε-DNP—Lys. Mild hydrolysis of **X** with dilute HCl gives a tripeptide **Y** and a tetrapeptide. Aspartic acid is cleaved from **Y** by BrCN.

On the basis of the foregoing evidence, and the fact that **S** fails to give a thiohydantoin derivative using the Edman degradation procedure, assign an amino acid sequence to **S**.

How is your answer modified by the further evidence that **S** has a molecular weight of about 1500, and that it can be polymerised by mild oxidation?

5. Substance **P** is a hormonal peptide that is produced by the hypothalamus gland of mammals. It has important pharmacological properties, and the sequence of bovine (i.e. from beef or cows) substance **P** was determined in 1979. The results are summarised below:

Starting with 180 kg of bovine hypothalami, the constituent peptides were extracted into solution, and purified by a combination of gel filtration, ion exchange chromatography, and paper electrophoresis. This yielded pure substance **P** (total weight 0.5 mg). Amino acid analysis was carried out after total acidic hydrolysis, giving the following results:

[Arg, Gly, Glu(2), Leu, Lys, Met, Phe(2), Pro(2)]

Enzymic hydrolysis revealed that substance **P** contained glutamine (but no glutamic acid). When substance **P** was hydrolysed for a limited period by the enzyme papain, certain of the amide bonds were totally or partially cleaved, generating six peptides; this meant that there were overlapping sequences resulting from the hydrolysis, although the specific sites attacked by papain were not known. These six fragments (F1–F6) were separated by paper electrophoresis; they were subjected to amino acid analysis after acidic hydrolysis, and the N-terminal residues were identified by formation of the dansyl derivatives (underlined below):

F1 [Arg, Glu, Lys, Pro(2)]   F4 [Gly, Phe(2)]

F2 [Arg, Glu(2), Lys, Pro(2)]   F5 [Leu, Met]

F3 [Glu, Gly, Phe(2)]   F6 [Leu, Met]

Fragment F1 was sequenced from the N-terminus using a combination of Edman and dansyl reagents (discussed further on page 157), sequentially giving the following dansyl derivatives:

Dns—Arg, Dns—Pro, Dns—Lys(Dns), Dns—Pro

Both substance **P** and fragment F6 were resistant to carboxypeptidase, but a single Edman degradation of F6 yielded Met—NH_2, which was identified by its electrophoretic mobility.

From these results, suggest a sequence for substance **P**.

6. The structure of tentoxin was determined by a combination of NMR and mass spectral data, and was confirmed by X-ray crystallography of the dihydro derivative (after catalytic hydrogenation of tentoxin).

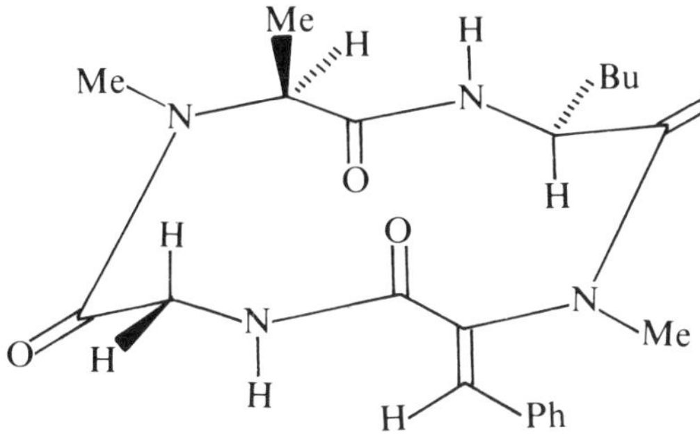

Tentoxin

(a) **Constituent amino acids**. Hydrogenation of tentoxin (H_2/Pd—C), followed by total acidic hydrolysis, gives four amino acids.

(i) Draw the amino acids.
(ii) Which three of the amino acids show identical chromatographic properties to appropriate DNA-encoded amino acids (or simple derivatives thereof)?
(iii) Which DNA-encoded amino acid might resemble the fourth amino acid from tentoxin (in physical and chemical properties), and how might its exact structure be determined?
(iv) What spectroscopic property of tentoxin would indicate the presence of the C=C double bond?

(b) **Sequence.**

(i) What evidence might indicate that tentoxin was cyclic?
(ii) How might mass spectrometry be used to sequence tentoxin?

(c) **Stereochemistry.**

(i) How might the absolute stereochemistry of one of the residues be determined?
(ii) How could NMR be used to reveal further information about the relative stereochemistry in tentoxin?

CHAPTER 6

Synthesis

Peptide synthesis is an essential tool for medicinal chemists. It allows them to:

  (i) Confirm the structure of a naturally occurring peptide (deduced from sequencing studies).

 (ii) Prepare relatively large amounts of rare peptides, for biological evaluation.

(iii) Make peptide analogues (hopefully with enhanced medicinal properties).

All of these considerations might apply to PENTIN, whose primary structure we elucidated at the end of Chapter 5:

$$Glp—Phe—Gly—Gly—Lys$$

So, how would we go about making PENTIN, starting from the readily available L-amino acids? In order to answer this question, let us start by considering the joining together of the first two residues of the C-terminus, to make the dipeptide Gly—Lys.

All that we need to do is to form an amide bond between the two residues—and we could manage this by taking glycine, converting it into the acid chloride, and then adding lysine.

 We would certainly get the desired dipeptide—plus about 100 by-products, including polymeric compounds, molecules formed by the amino group of glycine reacting with the acid chloride group (self-condensation), and products resulting from the side-chain of lysine forming an amide bond (instead of the α-

+ polymers of Gly/Lys

nitrogen reacting). In other words, there are three amino groups that could form an amide bond, but we only want a **specific** one (the lysine α-nitrogen) to react.

Therefore, we need to know how to protect the groups that we do not wish to react, remembering that they will need to be deprotected later on. Next, we will have to decide whether the use of acid chlorides is really the best way of coupling amino acids together. Finally, we must assess whether certain overall strategies (e.g. the order in which we condense the amino acids together) might be advantageous to the synthesis.

Once all these factors have been considered, then we can return to PENTIN itself, and devise a viable synthesis.

1 Protecting Groups

In order to form a dipeptide (such as Gly—Lys) from its constituent amino acids, we need to protect **all** of the functional groups except those that will be actually involved in the formation of the peptide bond; this includes amino and carboxylic

116

acid groups (for reasons that will be explained later), as well as any other reactive groups (such as nucleophilic alcohols or thiols, if they are present).

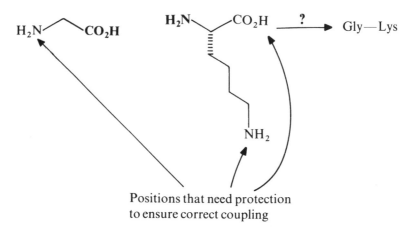

Positions that need protection
to ensure correct coupling

Moreover, we must choose the protecting groups so that they can be removed without rupturing the newly made peptide bond(s). First of all, let us consider the general tactics for protecting amines and carboxylic acids, before looking at the protecting groups themselves in more detail.

1.1 Amino Protection

The protection of amino groups is not straightforward, because the best way to render the nitrogen non-nucleophilic (and therefore unreactive) is to protect it as the amide! How could an amide-protecting group be removed without simultaneously cleaving any peptide bonds? The answer is to use **C-alkoxyamides (urethanes)**, which can be cleaved under milder conditions than normal amides (e.g. by careful acidic hydrolysis).

Nitrogen atoms are most readily protected as the alkoxycarbonyl derivatives:

$$R-O-\overset{\overset{\displaystyle O}{\|}}{C}-N\diagdown \quad \xrightarrow{\ H^{\oplus}(aq)\ } \quad R-OH + CO_2 + HN\diagdown$$

Alkoxycarbonyl protection of nitrogen

By changing the nature of the R group, deprotection can be achieved under a range of mild conditions.

1.2 Carboxylic Acid Protection

Carboxylic acids are easy to protect, by merely forming the corresponding ester. Esters are more easily hydrolysed than amides (e.g. mild alkaline conditions), so deprotection isn't usually a problem.

So we are now in a position to suggest a synthesis of the dipeptide Gly—Lys. By starting with the appropriately protected amino acids, we could carry out the synthesis as follows:

This tactic works perfectly well for our dipeptide, but we now have a fully deprotected compound. If we wished to continue with the synthesis of a longer peptide such as PENTIN, Glp—Phe—Gly—Gly—Lys, then we would need to selectively re-introduce new protecting groups—a difficult, time-consuming, and wasteful process. This problem could be overcome if we could just remove the α-amino-protecting group at the end of the synthesis.

By using two types of protecting groups, either of which can be selectively removed, peptide synthesis is simplified considerably. The use of complementary protecting groups is known as **orthogonal protection.**

1.3 Orthogonal Protection

There are many pairs of protecting groups that can be used in orthogonal protection, but one particular combination is used extremely widely in peptide synthesis: the benzyl and *t*-butyl type protecting groups.

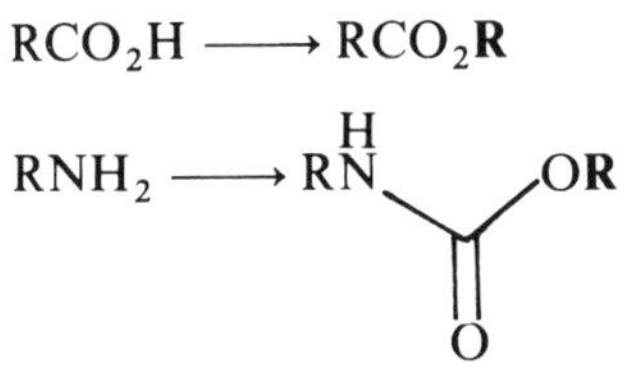

$$RCO_2H \longrightarrow RCO_2R$$

$$RNH_2 \longrightarrow RN\overset{H}{\underset{}{}}\overset{}{\underset{O}{C}}OR$$

Reagent R	TFA	H_2/Pd—C
R = CH$_2$Ph	×	✓
R = CMe$_3$ (Bu^t)	✓	×

× = stable   ($-CO_2CH_2Ph = Z$)
✓ = deprotection   ($-CO_2Bu^t = Boc$)

Benzyl/*t*-butyl orthogonal protection. The complementary nature of these protecting groups is shown above.
Benzyl can also be removed with **Na/NH$_3$(l)**, but **benzyl esters** yield the **amide** under these conditions.
All the above protecting groups can be simultaneously removed by treatment with **HBr/AcOH** or with **HF**.

By combining benzyl and *t*-butyl type protecting groups, we could easily prepare a protected Gly—Lys dipeptide that would be ready for further coupling reactions.

e.g.   **Boc—Gly—Lys—OCH$_2$Ph** ≡
         |
         Z

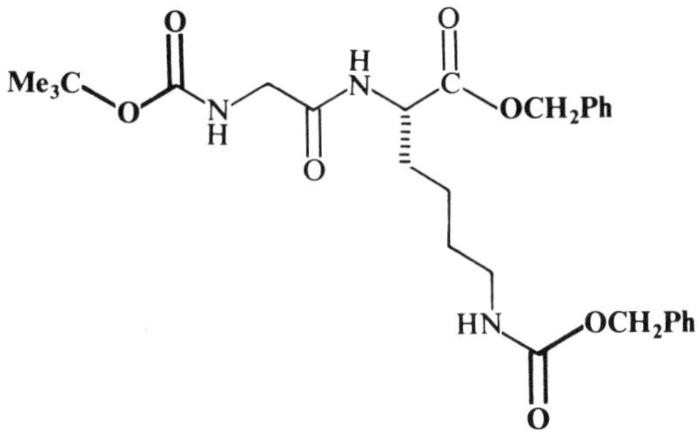

Treatment of this with TFA would deprotect only the α-amino terminus, whilst the C-terminus and the amino side-chain would remain protected. This free N-terminus could then be coupled to the next protected amino acid, and so on, until the desired sequence was obtained. For example, Glp—Phe—Gly—Gly—Lys could be prepared as follows:

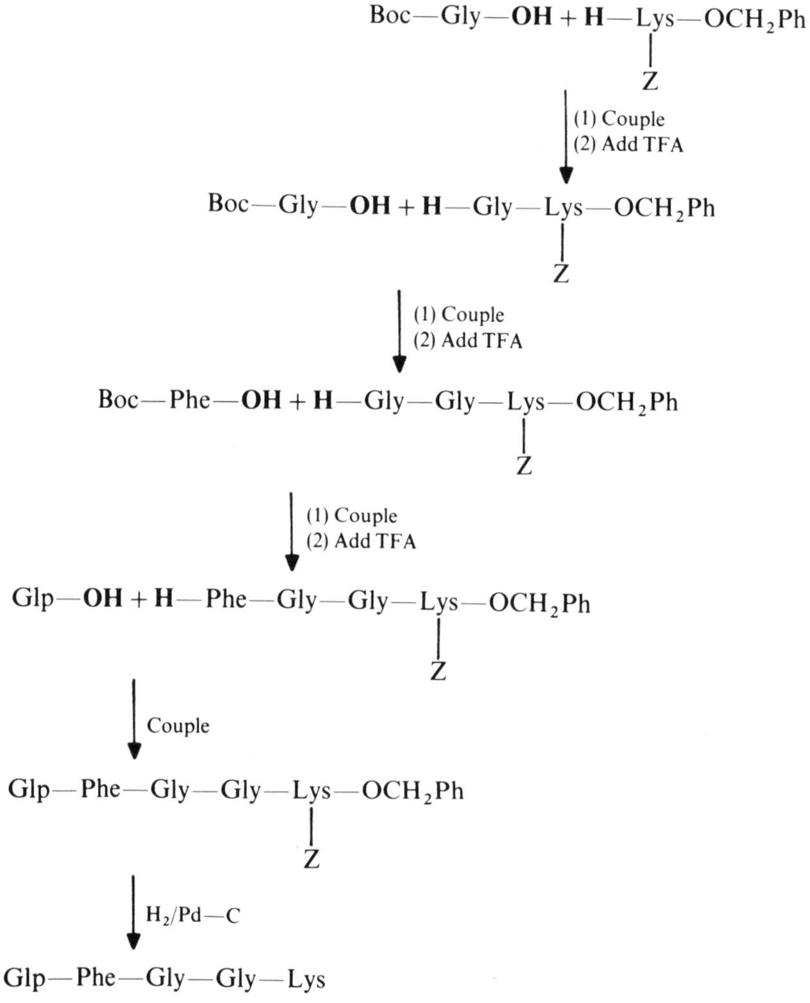

$$Boc—Gly—\textbf{OH} + \textbf{H}—Lys—OCH_2Ph$$
$$|$$
$$Z$$

(1) Couple
(2) Add TFA

$$Boc—Gly—\textbf{OH} + \textbf{H}—Gly—Lys—OCH_2Ph$$
$$|$$
$$Z$$

(1) Couple
(2) Add TFA

$$Boc—Phe—\textbf{OH} + \textbf{H}—Gly—Gly—Lys—OCH_2Ph$$
$$|$$
$$Z$$

(1) Couple
(2) Add TFA

$$Glp—\textbf{OH} + \textbf{H}—Phe—Gly—Gly—Lys—OCH_2Ph$$
$$|$$
$$Z$$

Couple

$$Glp—Phe—Gly—Gly—Lys—OCH_2Ph$$
$$|$$
$$Z$$

$H_2/Pd—C$

$$Glp—Phe—Gly—Gly—Lys$$

PENTIN

The details of this multi-step synthesis become rather long-winded. A common shorthand for peptide synthesis is shown below. Each residue is represented as a vertical line, and horizontal lines represent bonds; short diagonal lines show any side-chain protection, and reagents are given between structures.

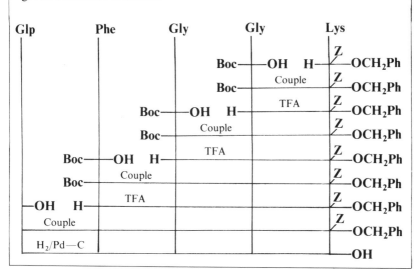

1.4 Protection of Other Groups

In the examples given above, only amino or carboxylic acid groups required protection. But side-chains often contain other functional groups that need protection during peptide coupling reactions; many protecting groups have been developed, and a small selection of the most widely used ones are given below:

Protected amino acid	Abbreviation	Removal
	$Arg(NO_2)$	$H_2/Pd—C$

Protected amino acid	Abbreviation	Removal
	Cys(**CH₂Ph**)	Na/NH₃(l)
	His(**Bom**)	H₂/Pd—C
	Ser(**CH₂Ph**)	HBr/AcOH or Na/NH₃(l)
	Tyr(**CH₂Ph**)	HBr/AcOH or Na/NH₃(l)

The side-chain protecting groups given above are all stable to treatment with TFA. All are removable by treatment with HF.

1.5 Introduction of Protecting Groups

So far, we have avoided the question of how to introduce the protecting groups. In fact, most peptide synthesis is carried out using amino acids which are bought with the protecting group(s) already attached. In other words, it is often cheapest and easiest to obtain them from commercial suppliers, who can make them in bulk.

We might need to protect an amino acid ourselves if we were trying to make a peptide with an unusual residue (i.e. the required **protected** amino acid was unavailable commercially), or if we needed large quantities of an expensive derivative. The simplest method of introducing each of the four commonest protecting groups is shown below:

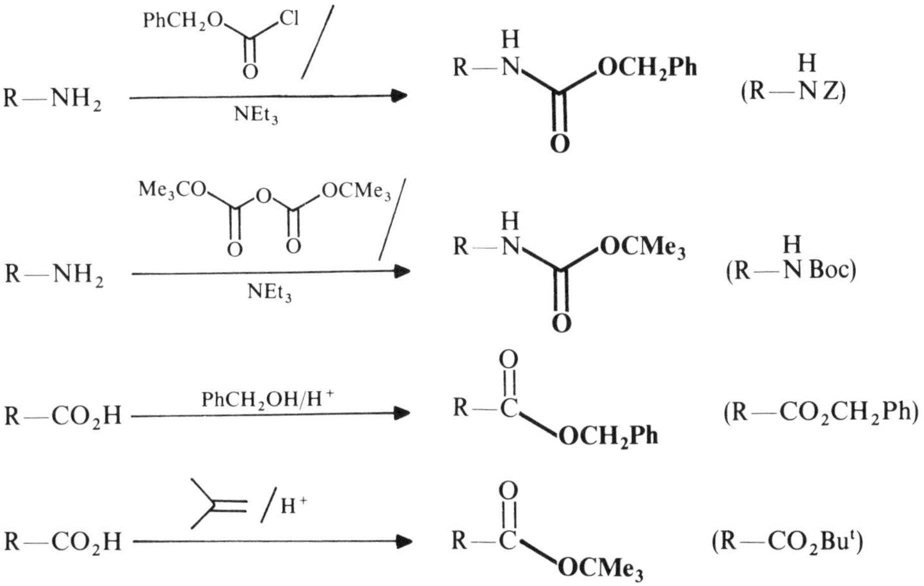

It should be apparent that protecting the **side-chain** of amino acids poses particular problems, because it is often necessary to differentiate between two carboxylic acid or amino groups.

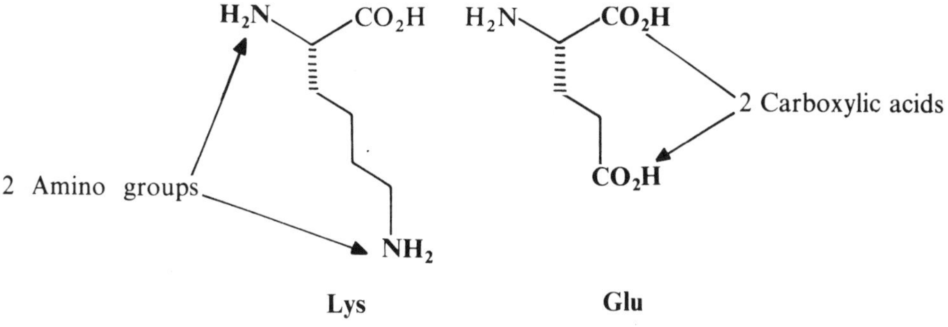

Amino acids that might require differential protection of α-groups and side-chains

124

For Boc—Lys(Z)—OH, the side-chain Z-protection can be introduced whilst the α-groups are chelated to a metal cation such as Cu^{2+} (see pages 27–28)

After removal of the cation, the α-amino group can be Boc-protected.

But selective protection procedures often need to be designed for a specific amino acid, and it is normal to look up such conversions in chemical journals— or simply buy the correctly protected amino acids.

The protecting groups described above would appear to meet all the requirements of the peptide chemist—easy to introduce, stable during coupling reactions, and readily removed under mild, selective conditions. But in fact, none of the protecting groups satisfy all these requirements, and dozens of new alternatives are devised every year. However, benzyl and *t*-butyl protection continue to be widely used in the field of peptide synthesis.

2 Principles of Peptide Coupling

First of all, let us consider the best way of forming the amide bond between Gly and Lys. Provided that we use suitably protected amino acids, we simply need to turn the OH of the carboxylic acid component into a good leaving group.

Boc—Gly—OH   Boc—Gly—X

When a peptide bond is formed between two amino acid residues, the **condensation** reaction is usually initiated by **activation** of the carboxylic acid component, and leads to the residues becoming **coupled**.

Activate   Coupled residues

One of the most widely used methods of activating carboxylic acids is to convert them into the corresponding acid chlorides; reaction with an amine forms an amide, which can be used in the identification of amines or carboxylic acids.

This approach works well for the formation of simple amides (as in the scheme above). But for peptide bond formation, problems occur if the carboxylic acid component is activated too strongly; in particular, a number of unwanted side reactions start to take place, of which the most serious is racemisation.

Racemisation is the loss of optical integrity at a chiral centre. For amino acids and peptides, this usually refers to the chiral α-carbon (which is L for DNA encoded amino acids) being converted into a D/L mixture.

If there is a carbonyl group adjacent to a chiral centre, then the system is often susceptible to racemisation. This could occur under either acidic or alkaline conditions:

For amino acids and peptides, the chiral α-carbons are necessarily adjacent to carbonyl groups, but racemisation is not usually a problem (except under extreme conditions). But when the carboxylic acid group of a protected amino acid is activated, then base-catalysed racemisation can be a serious problem.

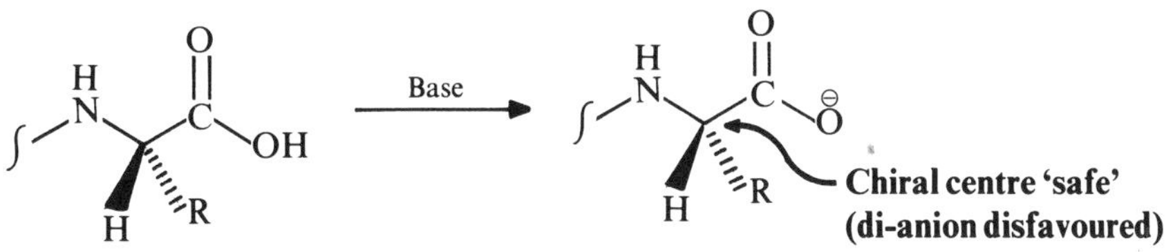

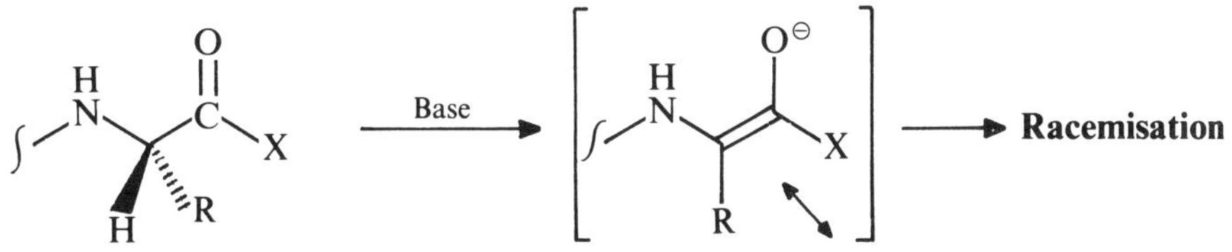

A free carboxylic acid suppresses base-catalysed racemisation of an adjacent chiral centre because (under basic conditions) it readily forms the carboxylate anion; removal of the α-proton would therefore generate the highly unstable di-anion. But activation of the carboxylic acid group not only makes removal of the α-proton viable, but also introduces an electron-withdrawing group (e.g. Cl in acid chlorides) which helps to stabilise the α-anion.

But for peptides (and amino acids with the amine protected as a simple amide), a much more important mechanism for racemisation can occur whenever the carboxylic acid OH group is replaced by a really good leaving group. Yet again, the mechanism involves a five-membered intermediate, called an oxazolone:

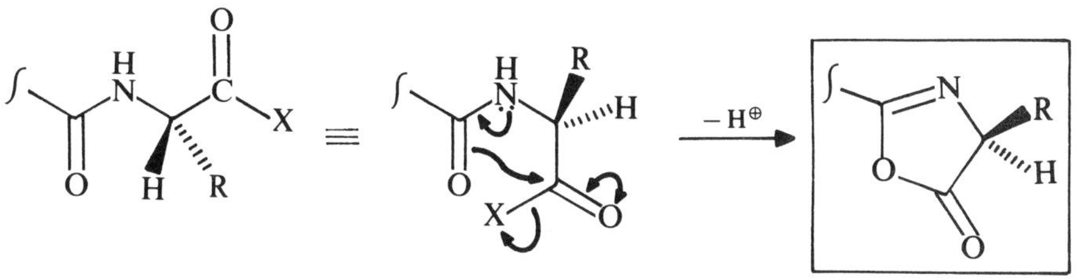

Oxazolones are fairly reactive molecules, and the peptide coupling could proceed smoothly enough if the required amino component were added:

It would not appear to matter, therefore, if some of the carboxylic acid derivative had been converted into the oxazolone; the new peptide bond could still be formed. However, the oxazolone is susceptible to racemisation under very mild conditions, particularly via base-catalysed deprotonation (yielding a **planar** aromatic ring with six π-electrons).

Doubly stabilised 'aromatic' anion

Therefore, any oxazolone formation could lead to racemisation of the chiral centre, meaning that the integrity of the newly made peptide would be in doubt.

There are four ways of dramatically reducing the degree of racemisation during peptide coupling reactions:

(i) Carry out couplings in which the activated carboxy residue is glycine (which is achiral) or proline (which fails to form the oxazolone).

(ii) Choose reaction conditions that minimise racemisation: low polarity solvent; neutral pH; low temperature.

(iii) Select the conditions for activating the carboxylic acid component with care (see below).

(iv) Use alkoxycarbonyl protected amino acids in the coupling reactions:

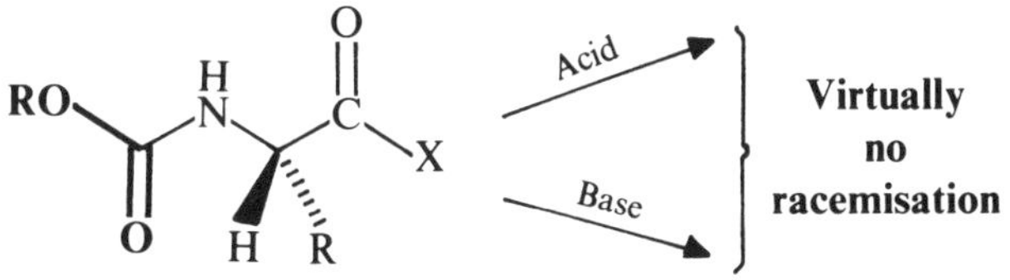

Incredibly, almost no racemisation is observed for these derivatives, except under very harsh conditions. (It is believed that the corresponding alkoxy-oxazolones are difficult to deprotonate at the α-carbon.)

> Alkoxycarbonyl protected amino acids can be 'activated', and then coupled to an appropriate amino component, with almost no risk of racemisation occurring.

The method of activation is an extremely important decision for the peptide chemist. On the one hand, activation via a really good leaving group generally leads to rapid and efficient coupling, but racemisation (and other side reactions) can be a problem. On the other hand, milder activating procedures mean longer reaction times (and perhaps incomplete coupling), but side reactions are limited. As a result of this dilemma, a very wide range of coupling methods have been devised, each with certain advantages and disadvantages. Just a few of these have been used really extensively by peptide chemists, and four of them are discussed in detail here.

3. Peptide Coupling Reagents

3.1 Dicyclohexylcarbodiimide

Despite the long name for this reagent (which is usually abbreviated to DCCD or DCC) the chemistry is really quite simple (see page 24). The important part is the diimide moiety, —N=C=N—, which contains a central carbon atom that is very electron deficient, and is therefore readily attacked by nucleophiles. So if DCCD is added to a carboxylic acid, the following reaction occurs:

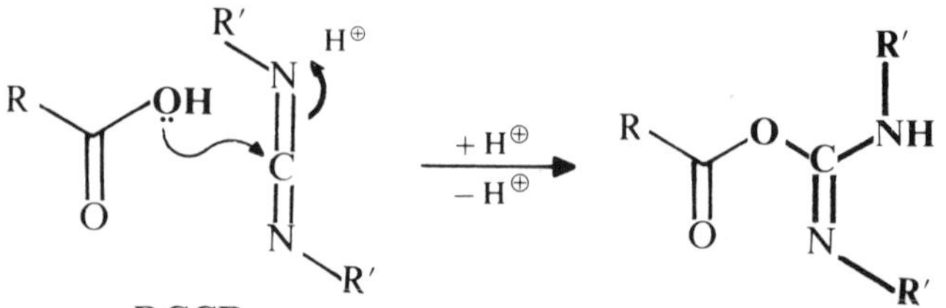

DCCD

The carbonyl group of the carboxylic acid derivative is now quite reactive, and the addition of an amine results in the formation of an amide bond:

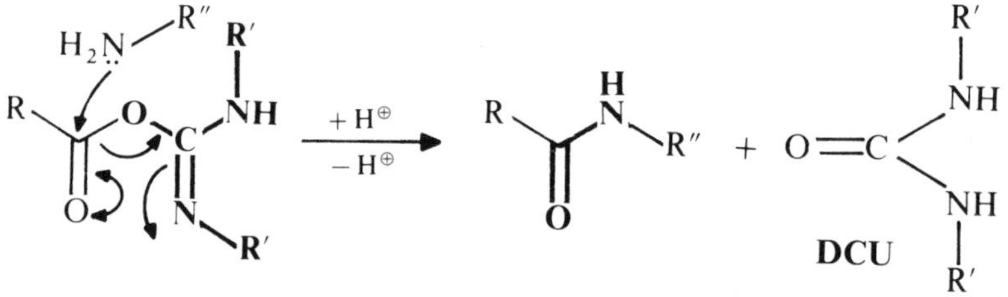

DCU

The DCCD finally ends up as dicyclohexylurea (DCU), which is a particularly stable molecule whose formation is a driving force for the reaction. It would be logical to suppose that the amino component must be added after the carboxylic acid has been 'activated' by DCCD, or else the nucleophilic amine might itself attack the DCCD. However, reaction of DCCD with an amine cannot lead to the formation of the stable DCU, and a series of equilibria are set up; the result is that DCCD can be added directly to the mixture of amine and carboxylic acid, and peptide bond formation proceeds smoothly and rapidly (usually less than 1 hour) using this procedure.

This so-called 'one-pot' procedure, in which acid, amine, and DCCD are just mixed without the separate formation of an 'activated' acid derivative, is particularly straightforward to carry out experimentally. The DCU formed in the reaction is relatively insoluble in most organic solvents, so a simple filtration step can leave almost pure coupled peptide in solution—if you're lucky! There are endless variations on the use of DCCD, in which the aim is to reduce further the already low degree of racemisation that is sometimes observed (e.g. see active ester couplings below). However, if the acid and DCCD are mixed **before** the amino component is added, then the coupling is still successful, but takes place via the anhydride, as shown below.

The use of **dicyclohexylcarbodiimide (DCCD)** allows a carboxylic acid and an amine to be coupled in a 'one-pot' reaction; it is the simplest method of peptide coupling.

3.2 Anhydride Couplings

If a carboxylic acid is stirred with DCCD ($\frac{1}{2}$ mole equivalent) for a few minutes, then a white precipitate of DCU appears, even before any amino component is added. This is caused by the self-condensation of the acid, forming a **symmetrical anhydride** which can be readily isolated.

Far from being a nuisance, the anhydride can then be used itself for the desired peptide coupling reaction, simply by mixing it with the appropriate amino component.

This procedure is very widely used in peptide synthesis, not only because the reaction is usually very clean (no reactive compounds such as DCCD present for the actual coupling step), but also because it takes place conveniently rapidly (usually within 1 hour). There is one major disadvantage, however; half of the original carboxylic acid component (often an expensive protected amino acid) is necessarily wasted during the reaction, ending up as carboxylate during the coupling step. Recovery of the carboxylate is rarely practicable, but other anhydrides have been devised in an attempt to overcome this problem.

For example, **mixed anhydrides** have been developed, in which only one of the carbonyl groups is susceptible to nucleophilic attack by amines.

More stable Bulky Bu^t group
C=O group

The pivolyl mixed anhydride is prepared by reaction of the carboxylic acid component with (the relatively cheap) pivolyl chloride (Bu^{t}COCl). When the requisite amine is added to the resulting anhydride, attack at the pivolyl carbonyl group is negligible, for steric and inductive reasons; the desired peptide bond is therefore formed, and unnecessary loss of the carboxylic acid component is avoided.

The phosphite derivatives shown below are another type of mixed anhydride. Again, when an amine is added, attack occurs almost exclusively at the desired carbonyl group:

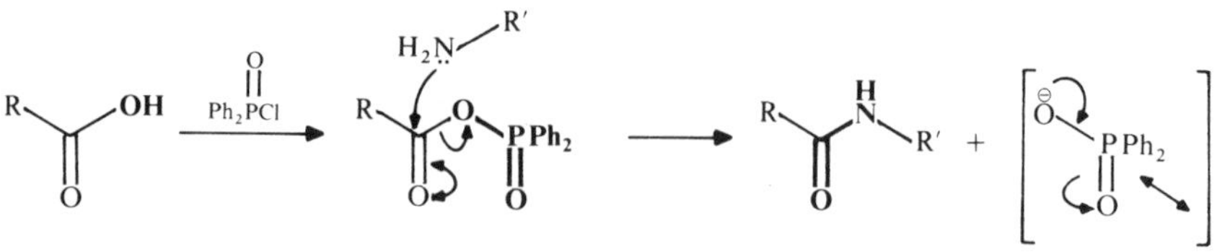

Nevertheless the **symmetrical anhydride** method has now become one of the most widely used procedures for coupling one residue at a time.

Symmetrical anhydrides (formed from an acid + DCCD) react with amines to give amides that are usually very pure, but this method of peptide coupling does involve the wastage of half of the carboxylic acid component.

3.3 Active Ester Couplings

If a carboxylic acid is converted into its methyl ester before heating it with an amine, then the corresponding amide is formed. This reaction is so sluggish, however, that it is not a practicable synthetic method for the peptide chemist. But if a different ester is used, in which the stability of the leaving group is improved, then a smooth coupling reaction can take place.

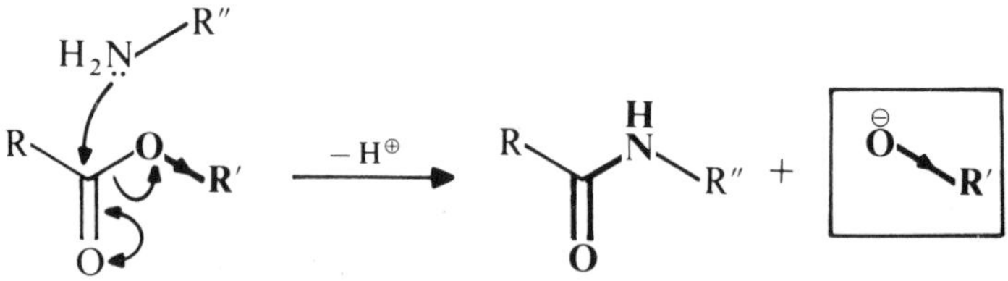

In other words, even simple esters will undergo effective peptide coupling with amines, provided that the $^-$OR leaving group can be stabilised in some way. This could be achieved via either mesomeric or inductive effects, and there are excellent examples of both:

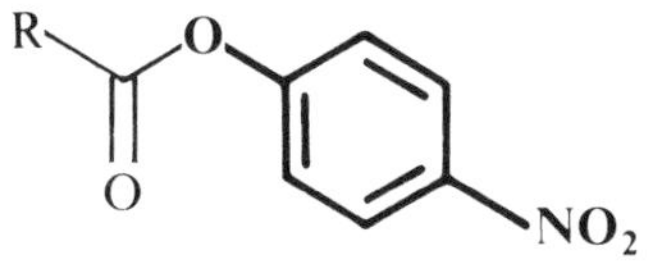
R
O
O
NO_2

p-Nitrophenyl esters
(*p*NP esters)

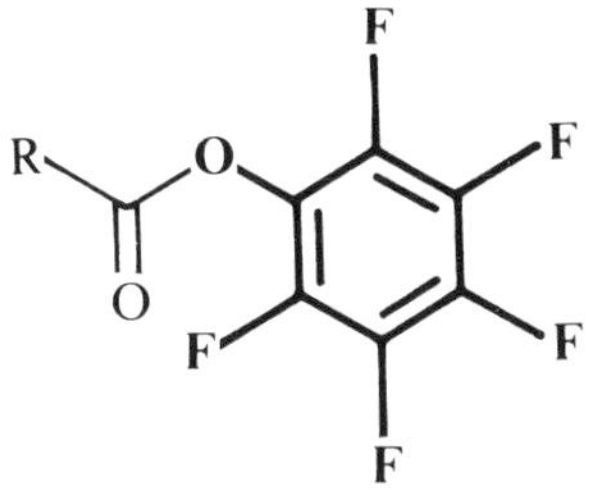
R
O
O
F
F
F
F
F

Pentafluorophenyl esters
(PFP esters)

The classical *p*NP esters are rather slow to react, requiring perhaps 12 hours for complete coupling, but the more recent PFP esters usually react within 1 hour. Both reactions tend to give very pure peptide products, although the actual formation of the active esters is sometimes a bit messy—often accomplished by reacting the acid with DCCD and the appropriate alcohol (see page 24).

Some active esters can be formed *in situ* during coupling reactions, and this is usually employed as a method of improving the cleanness of DCCD couplings:

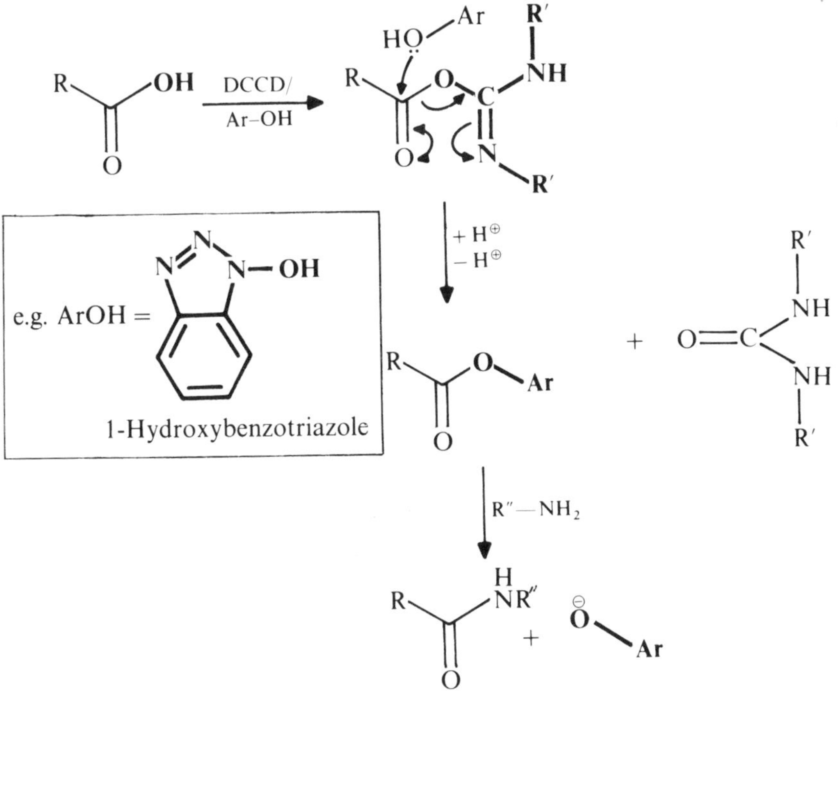
DCCD/
Ar–OH
+ H⊕
– H⊕
e.g. ArOH =
1-Hydroxybenzotriazole
R″—NH_2

In situ **active ester coupling.** For example, 1-hydroxybenzotriazole (Ar—OH) and DCCD are reacted with the carboxylic acid component in a 'pre-activation' step; subsequent addition of the amine leads to clean formation of the desired amide.

Active ester couplings are increasing in popularity for a couple of reasons. Firstly, the active esters of many protected amino acids are now commercially available. Secondly, it is sometimes possible to monitor coupling reactions by the UV absorbance of the anion generated from active ester couplings.

When **active esters** are reacted with amines, amides of high purity are usually obtained. The most widely used active esters are those of 4-nitrophenol (*p*NP) or pentafluorophenol (PFP), but active esters can be difficult to prepare (or expensive to buy).

3.4 Azide Couplings

Acid azides can be readily formed from carboxylic esters, by reaction with hydrazine (forming the acid hydrazide) followed by nitrous acid (a source of $[NO]^+$):

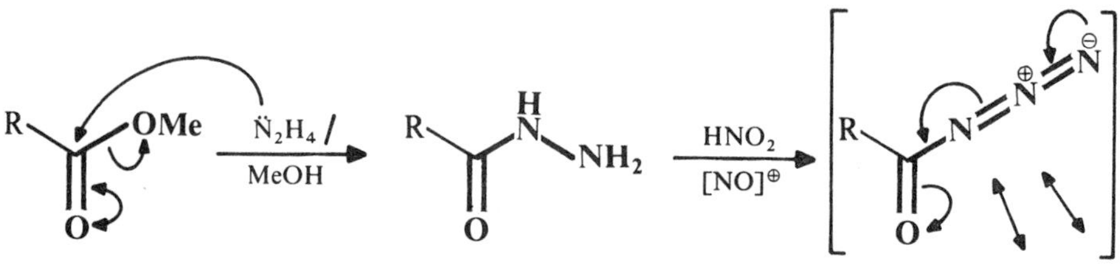

Azide formation. Using N_2H_4 in MeOH, esters can be converted into hydrazides without any other significant reactions taking place.

$RCON_3$; resonance stabilised acid azide

The azide anion (N_3^-) is a moderately good leaving group. So reaction with amines takes place over 10 hours or so, to give the corresponding amide.

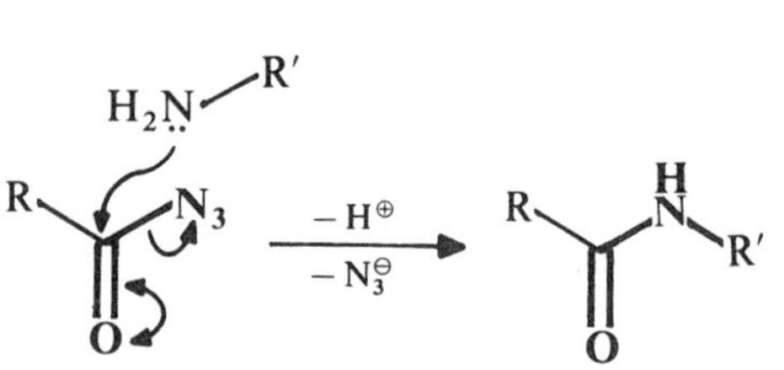

The azide method is not usually the coupling procedure of choice, because the acid azides can undergo a number of side reactions (e.g. Curtius rearrangement), and because the reaction is rather slow—but the azides are extremely resilient to racemisation. Therefore, they are particularly valuable for couplings in which there is a really serious risk of racemisation, particularly when two peptide fragments are being condensed.

The **azide** method of peptide coupling involves the use of an acid azide and an amino component. The reaction is only used when racemisation is a serious problem, particularly in the coupling of two peptide fragments.

The four main methods of peptide coupling are summarised in Table 6.1 on the next page.

4   Strategy

We now know how to couple amino acids together, and also how to use protecting groups so that only the specific peptide bond that we require is formed. So we ought to be able to synthesise a peptide such as PENTIN (Glp—Phe—Gly—Gly—Lys). But should we conduct the synthesis by adding on just one amino acid at a time, or should we construct our target molecule in chunks? And are there other factors that might influence our design of a synthetic route? There are four main considerations, although they are interdependent to some extent.

4.1   Linear vs Convergent Synthesis

A peptide synthesis is said to be **linear** if the target molecule is built up by the stepwise coupling of one residue at a time. A **convergent** peptide synthesis involves the separate synthesis of peptide fragments (containing two or more residues), which are then coupled to give the target peptide.

Consider the synthesis of an octapeptide, whose amino acid sequence we shall designate as:

$$A-B-C-D-E-F-G-H$$

Suppose that we could guarantee a 90% yield at each coupling step (which isn't

Table 6.1. The main methods of peptide coupling.

Method	Activated form: R—CO—X	Advantages	Disadvantages
1 DCCD		Simple	Some side reactions
2 Anhydride		Very clean reaction	Wasteful, and quite expensive
3 Active ester		Very clean reaction	Active esters expensive, or tricky to prepare
4 Azide		No racemisation in fragment couplings	Several side reactions

bad by most people's standards!), then how would a convergent route compare with a linear one, in terms of overall yield?

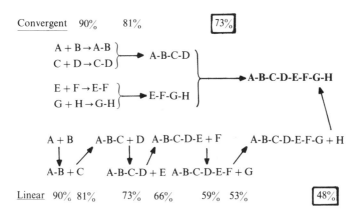

Looking at the final yields, there is clearly an enormous advantage with conducting a convergent synthesis. Moreover, although both of the routes need seven steps, the reactions using the linear approach require the commitment of sequentially bigger and bigger peptides to yet another coupling reaction; using the convergent approach, most of the chemistry is carried out using relatively small fragments. So, considering both overall yield, and the practicalities of the chemistry involved, it would appear that the convergent approach is the better strategy.

> **Convergent** peptide synthesis generally gives higher overall yields than **linear** synthesis, and involves the manipulation of small, easily handled peptide fragments for most of the steps.

4.2 Which Fragment Couplings?

If we consider that the convergent approach is the best tactic for peptide synthesis, then a pentapeptide such as PENTIN could be prepared in several different ways (see p. 138).

Is one of these strategies significantly better than the others? The answer is 'yes'; route D should involve an easier fragment coupling because **racemisation** will not be a problem. Even using the simple DCCD coupling procedure, the activated component will be free from racemisation because glycine is achiral.

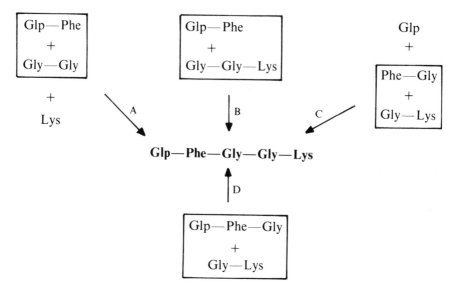

Possible convergent routes to PENTIN

> **Convergent synthesis of PENTIN.** Using **route D**, the key fragment coupling involves activation of achiral glycine. However the tripeptide (and dipeptide) precursors can only be prepared by linear synthesis.

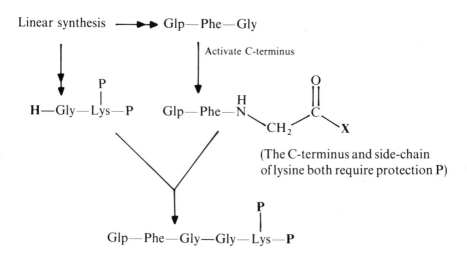

(The C-terminus and side-chain of lysine both require protection P)

In selecting **fragment couplings**, it is preferable to choose carboxylic acid components that are not susceptible to racemisation; i.e. **C-terminal glycine or proline** residues (see page 128); then the simple DCCD coupling procedure can be adopted. If C-terminal Gly or Pro cannot be used in fragment couplings, then the azide coupling procedure is often chosen (see pages 134–135).

4.3 Should Peptide Synthesis Start from the N- or C-terminus?

The formation of peptides almost always requires some linear synthesis, even if the overall approach is convergent. For example, if we used route D for the synthesis of PENTIN (see above), then we would need a tripeptide fragment— which would need to be made by linear synthesis.

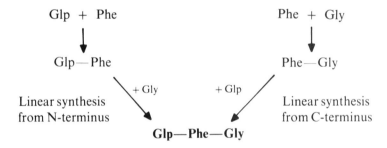

Is it better to make this Glp—Phe—Gly tripeptide by building it up 'from left to right' or 'from right to left'? The answer is 'from right to left', because racemisation can be minimised. You may recall that, when the α-nitrogen of an amino acid is protected with the standard alkoxycarbonyl group, almost no racemisation takes place when the carboxylic acid group is activated (see page 129). This means that any method of activation can be used, without there being any serious risk of racemisation occurring.

If the synthesis had been started from the N-terminus, then we would have had to activate Glp—Phe; the α-carbon of Phe might have undergone some racemisation, unless the rather messy azide coupling procedure had been employed. Moreover, it is usually preferable to activate a simple amino acid (and then add a more complex amino component) rather than to activate a more complex peptide (and then add a simple amino component).

For **linear peptide synthesis**, it is better to **start from the C-terminus** rather than from the N-terminus. This is because alkoxycarbonyl-protected amino acids can be readily activated **without serious risk of racemisation.**

$$\textbf{Boc}\text{---Phe}\text{---}\textbf{OH} + \textbf{H}\text{---Gly}\text{---}\textbf{OCH}_2\textbf{Ph}$$

$$\downarrow \text{DCCD}$$

$$\textbf{Boc}\text{---Phe}\text{---Gly}\text{---}\textbf{OCH}_2\textbf{Ph}$$

$$\downarrow \text{TFA}$$

$$\textbf{H}\text{---Phe}\text{---Gly}\text{---}\textbf{OCH}_2\textbf{Ph}$$

$$\downarrow \text{Glp/DCCD}$$

$$\text{Glp}\text{---Phe}\text{---Gly}\text{---}\textbf{OCH}_2\textbf{Ph}$$

$$\downarrow \text{H}_2/\text{Pd}\text{---C}$$

$$\boxed{\text{Glp}\text{---Phe}\text{---Gly}}$$

Using the considerations above, it is quite easy to devise a synthesis of PENTIN which is both **convergent** and virtually **racemisation free** at every step.

Having produced such convincing arguments for conducting a convergent synthesis, it should be something of a shock to discover that the vast majority of peptide synthesis is now conducted in a linear fashion. This is because of the advent of solid-phase peptide synthesis.

4.4 Solid-phase Peptide Synthesis

One of the biggest problems in peptide synthesis is not so much the actual protection/deprotection steps or the coupling reactions themselves, but the need to purify the peptide at every stage to avoid impurities accumulating. This is time consuming, and often far from easy, even with modern purification methods. Although the reactions look clean enough on paper, the steady build-up of by-products in the multi-step synthesis of peptides has been a major problem.

One way of simplifying the purification considerably would be to attach the growing peptide to some functional group 'handle' which could be isolated (with its attached peptide) from impurities in the reaction. The obvious position at which to attach such a 'handle' would be the carboxy terminus, because peptides are usually built up from that end (see above). In the 1960s, R. B. Merrifield developed the technique of attaching the C-terminus to an insoluble polymeric support. This tactic was to revolutionise peptide synthesis, and earned Merrifield the Nobel Prize for Chemistry in 1986.

Merrifield used a specially cross-linked polystyrene support, and then reacted this with chloromethoxymethane in the presence of Lewis acid catalysis. This Friedel–Crafts reaction yielded an insoluble polymer, with the potential to form a benzyl ester type linkage to an appropriate amino acid derivative:

These polymers usually come in the form of small beads, and the attached amino acid reacts exactly as if it were a simple benzyl ester; however, because of the cross-linking, the polymer (plus attached amino acid) is insoluble in organic solvents.

The aim is to ensure that the growing peptide is the only organic molecule attached to the resin. Any impurities can then be removed by simply washing the polymer—a very quick and easy process.

If this tactic is to work successfully, then we must guarantee virtually 100% deprotection/coupling at every step, or else peptides with missing residues will also be attached to the resin. Coupling methods that ensure complete reaction within 1 hour are usually employed, and a large excess of the activated amino acid (say, 3–5 mole equivalents) is normally added.

Merrifield's original combination of protection and activation are still widely used: α-nitrogen protection with Boc, activation with DCCD, and benzyl-type side-chain protection. At the end of the synthesis, cleavage of the benzyl ester type linkage to the resin can be accomplished at the same time as removing all the protecting groups, using HBr/AcOH.

So the following cycle of reactions is carried out during peptide synthesis:

More recently, other combinations of polymeric support, C-terminal linkage, protection and activation have been developed, as exemplified below.

(i) Polymeric support

Radical polymerisation of dimethylacrylamide $\left(\text{Me}_2\text{N}\overset{O}{\diagdown}\right)$ is carried out in

the presence of a functionalised derivative $\left(\text{MeO}_2\text{C}\diagdown\overset{O}{\underset{Me}{N}}\right)$ and a

cross-linking agent $\left(\diagup\overset{O}{\underset{H}{N}}\diagup\overset{O}{\underset{H}{N}}\right)$. The resulting polymer is permeable

to solvents like DMF (Me$_2$N—CHO), can be readily derivatised *via* the ester group, and can also be supported on Kieselguhr (an inorganic clay), to allow 'continuous flow' peptide synthesis (in which the reagents and solvents flow through a column containing the resin).

(ii) Peptide-polymer linkage

A two-step modification of the methyl ester group gives a hydroxy derivative, to which the C-terminal amino acid can be attached by a simple ester bond

$(\text{RCO}_2\text{H} + \text{HO}\text{—R}' \xrightarrow{\text{DCCD}} \text{RCO}_2\text{R}')$. The linkage is acid labile, so the final

peptide can be released by treatment with TFA.

144

(iii) Protection

Fmoc

The **Fmoc** protecting group is readily removed by mild base (20% piperidine in DMF is usually employed). By using N^α-**Fmoc** protection, the growing peptide need only come in contact with mild reagents until the final deprotection. If t-butyl side chain protection is employed, then final treatment with TFA liberates the fully deprotected peptide from the resin.

(iv) Activation

ODhbt

The **Dhbt** esters couple fairly rapidly to free amino groups (about 30 minutes). Whilst coupling is taking place, a yellow colour is produced on the resin, which fades when coupling is complete. This allows the peptide bond formation to be monitored spectrophotometrically.

Four factors have ensured that the solid-phase approach now dominates peptide synthesis:

(i) The syntheses are very fast to carry out; 5–10 residues can easily be coupled together in one day—an octapeptide might take a month or more to synthesise using conventional solution chemistry.
(ii) The steps are simple and repetitive, and it has been possible to optimise the reactions to almost 100%. So yield and purity are quite good for peptides up to about 15 residues.
(iii) Purification by HPLC (particularly reversed phase) has enabled peptides from solid-phase synthesis to be isolated rapidly in a highly pure state.
(iv) Because the method is so simple and repetitive, it has been possible to fully automate it. Automatic peptide synthesisers are expensive, but they are good value for pharmaceutical firms, who can operate them for 24 hours a day.

Solid-phase peptide synthesis does have a few drawbacks, however. In particular:

(i) Impurities can start to build up in big peptides (because the growing peptide is not purified rigorously at every stage), although 15–20 residues is usually alright.
(ii) Sometimes a coupling can be unexpectedly difficult, probably due to the folding of the growing peptide. This can be hard to detect during solid-phase peptide synthesis.
(iii) There is quite a serious limit on the amount of peptide that can be prepared by solid-phase techniques. The resins cannot be loaded too heavily with the first residue, or inefficient couplings are observed: 0.5 g of resin (which is conveniently handled in most laboratories) would only yield about 75 mg of a pentapeptide (which is plenty for simple biological testing, but could not possibly meet medical needs).
(iv) The resins, protected amino acids, reagents, and solvents for solid-phase peptide synthesis are all quite expensive.

Not surprisingly, all of the problems outlined above are receiving a lot of attention from research chemists, and solutions are being discovered. But the solid-phase approach already dominates the field of peptide synthesis, and it will undoubtedly become even more important in the future.

> **Solid-phase peptide synthesis** immobilises the growing peptide on an insoluble polymer, allowing easy purification after each coupling. It is a fast and reliable way of making small quantities (e.g. 10–100 mg) of peptides for biological testing. Larger quantities of peptides are more likely to be prepared using **convergent solution chemistry.**

5 Summary

We are now in a good position to attempt a synthesis of PENTIN. But we must decide the **quantity** of peptide that we need, as this will affect our overall synthetic strategy.

Glp—Phe—Gly—Gly—Lys

Initially, we will only require a few milligrams of synthetic PENTIN—just enough to confirm that our conclusions from the sequencing studies were correct, and to carry out some simple biological tests. In the longer run, we are likely to need larger quantities of PENTIN—perhaps grams of the material, if it is being tested as a new pharmaceutical; and tens or hundreds of grams if it actually becomes accepted for medicinal use.

5.1 Milligrams of PENTIN

In order to confirm the structure of PENTIN, we only need to synthesise milligram quantities, but we would like to get the results very quickly—perhaps in two or three days.

146

The synthesis of a pentapeptide requires about 12 separate synthetic steps (including activation, coupling, and deprotection processes), but using **solid-phase peptide synthesis**, we could achieve this in 6–8 hours! We would only generate perhaps 20–30 mg of PENTIN, but this would be ample for the initial studies.

If we chose Merrifield's procedure, then we would need to order the following amino acids:

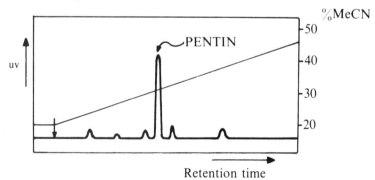

| Glp | Boc—Phe | Boc—Gly | N^α-Boc, N^ϵ-Z—Lys |

Attaching the first residue to the resin is a relatively sluggish step—although we could buy a resin with the first residue already attached. Thereafter, the α-amino groups can be liberated by treatment with TFA, and the next residue coupled by the addition of DCCD (negligible racemisation of N^α-alkoxycarbonyl amino acids)— see Scheme on page 147.

Final deprotections and removal from the resin can **all** be achieved by treatment with **HBr/AcOH**, to give crude synthetic PENTIN; 0.25 g of initial resin would yield about 40 mg.

In order to purify the peptide, we would use **reversed phase** HPLC, hoping to obtain the following trace:

Retention time

Isolation of the major peak should yield synthetic PENTIN, whose composition could be checked by **amino acid analysis**. We could then check that naturally occurring PENTIN was identical with the synthetic material—HPLC retention time would be the first criterion, followed by biological assays.

If PENTIN did prove to have promising biological properties, then we would certainly need more of it for further studies.

$$\text{Glp—Phe—Gly—Gly—Lys—OH}$$

PENTIN

5.2 Grams of PENTIN

For these sort of quantities, we would turn to **solution peptide synthesis**. We would probably use a **convergent** approach, although we might well employ **orthogonal** *t*-butyl/benzyl protection (and so employ the same protected amino acids as in the solid-phase method above).

We would choose our fragment coupling carefully, in order to minimise the risks of racemisation, so that an overall synthetic sequence might be as follows:

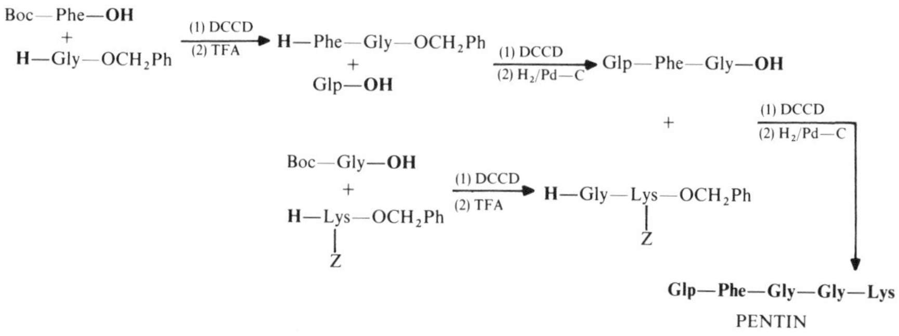

It would probably take a few weeks to get all the reactions working smoothly, but this method would allow us to prepare grams of PENTIN, and to scale up to larger quantities if necessary.

So we could make synthetic PENTIN! We could confirm the structure of the naturally occurring material in days using the solid-phase approach, and prepare much larger quantities for major testing in a few weeks.

But undoubtedly, in conjunction with any biological testing of PENTIN, we would look at its analogues to see whether any of them had better medicinal properties than PENTIN itself. Studies of this type are central to the development of most new drugs. Fortunately, the medicinal chemist can prepare a range of analogues for simple bio-assays quite quickly, using solid-phase peptide synthesis. Many considerations can guide the chemist to the most likely compounds for testing, and the following factors would need to be considered:

Potency of the drug
Stability in the body
Rate of absorption
Side-effects

There are many approaches to rational drug design, and we will not consider them in detail here. But, in the end, it is only by synthesising the most promising structures that the medicinal properties can be evaluated.

Our story of PENTIN is a fictitious one, but it is not unrealistic. In the 1970s, a hormonal peptide called LH-RH was isolated from pigs. The elucidation of its structure, which was only confirmed by total synthesis, was an outstanding achievement for the peptide chemists involved. LH-RH was found to possess extensive biological properties, and an analogue of LH-RH called Zoladex[R] was released in 1989, for use in the treatment of cancer. The story of LH-RH, and its modification to give an anti-tumour drug, is told in Chapter 7.

Further Reading

General Texts

The Chemical Synthesis of Peptides, John Jones, Clarendon Press, 1991. A good coverage of the principles of solution and solid phase peptide synthesis, including many case studies.

Peptide Synthesis, 2nd Edition, M. Bodanszky, Y. S. Klausner, and M. A. Ondetti, Wiley, 1976. A classic general coverage of the main aspects of peptide synthesis.

The Peptides; Analysis, Synthesis, Biology, Volumes 1, 2, and 3 (E. Gross and J. Meienhofer, Eds.), Academic Press, 1979–81.

Very thorough coverage of:

    Vol. 1—Peptide coupling procedures

    Vol. 2—Solid-phase peptide synthesis (Ch. 1, pp. 1–284)

    Vol. 3—Protecting groups in peptide synthesis

The Proteins; Composition, Structure, and Function, Volume 2, 3rd Edition (H. Neurath and R. L. Hill, Eds.), Academic Press, 1976. Valuable detailed coverage, in this major series.

Solution-phase Synthesis

Principles of Peptide Synthesis, M. Bodanszky, Springer-Verlag, 1984. Concentrates on the principles of solution-phase peptide synthesis; goes hand-in-hand with:

The Practice of Peptide Synthesis, M. Bodanszky and A. Bodanszky, Springer-Verlag, 1984. Comprises experimental details for the main procedures in use (up to 1984) for solution-phase peptide synthesis.

Solid-phase Synthesis

Solid Phase Peptide Synthesis, 2nd Edition, J. M. Stewart and J. D. Young, Pierce, 1984. A very useful text.

R. B. Merrifield, *Angew. Chem. (Int. Ed.)*, 1985, **24**, 799. Merrifield's Nobel Prize Lecture—an interesting read.

The Peptides; Analysis, Synthesis, Biology, Volume 9 (S. Udenfriend and J. Meienhofer, Eds.), Chapter 1, pp. 1–38, Academic Press, 1987. R. C. Sheppard discusses the use of the Fmoc protecting group; for his work on Fmoc PFP esters, see *Tetrahedron*, 1988, **44**, 843 and 859. For Fmoc Dhbt esters, and fully automatic peptide synthesis, see *J. Chem. Soc., Perkin I*, 1988, 2887 and 2895).

Solid Phase Peptide Synthesis: A Practical Approach, E. Atherton and R. C. Sheppard, IRL Press, 1989. A very recent practical book in this important area.

Questions

1. List the chemicals (including derivatised resin, activating/deprotecting reagents, and protected amino acids) that would be needed in order to synthesise each of the following peptides using soild-phase methodology:
   (a) Ala—Pro—Gly
   (b) Leu—Lys—Phe—NH$_2$
   (c) Glu—His—Ser—OMe
   (d) H – Cys—Gly—Gly—Gly—Gly—Cys—OH

2. Draw the structure of the following tetrapeptide, and outline the steps required to prepare it using solid-phase techniques.
   $$\text{Pro—Lys—(D)-Phe}$$
   $$\text{Ac—Ala}\underset{}{\overline{}}$$

3. Deduce the structure of the peptide (Z) resulting from the following sequence of reactions and explain the major peaks observed in the FAB mass spectrum. The sequence of reactions start from a derivatised polymer (see page 143), using the solid-phase method (washing steps omitted); why can the peptide be cleaved so readily by the addition of strong acid (e.g. TFA)?

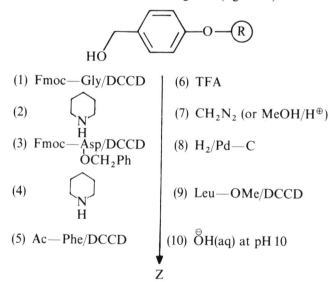

(1) Fmoc—Gly/DCCD	(6) TFA
(2) [piperidine, N—H]	(7) CH$_2$N$_2$ (or MeOH/H$^{\oplus}$)
(3) Fmoc—Asp/DCCD, OCH$_2$Ph	(8) H$_2$/Pd—C
(4) [piperidine, N—H]	(9) Leu—OMe/DCCD
(5) Ac—Phe/DCCD	(10) $\overset{\ominus}{O}$H(aq) at pH 10

Z

FAB MS (Z$^+$ ion) gives major peaks *m/z*: 493, 418, 362, 190, and 43.

4. This cyclic hexapeptide was required for conformational studies:

$$\begin{array}{ccc} \text{Phe} \longrightarrow \text{Gly} \longrightarrow \text{Phe} \\ \uparrow \qquad\qquad \downarrow \\ \text{Phe} \longleftarrow \text{Gly} \longleftarrow \text{Phe} \end{array}$$

The cyclisation step was expected to be low yielding, so about 1 g of the linear precursor was prepared. Outline a synthesis of a suitable linear precursor. (NB. Solution methods will be required for such quantities, and convergent syntheses can be very efficient in time and yield.) What conditions would you propose for the attempted cyclisation?

5. Enkephalin (A) is a naturally occurring peptide that interacts with receptors in the brain to elicit pain relief. The cyclic peptide (B) is a synthetic analogue that is even more effective than the natural compound.

Tyr—Gly—Gly—Phe—Leu
A

B

(a) Briefly describe the structural relationship between A and B.
(b) Outline a chemical synthesis of Boc—Tyr—Glu—Gly—Phe.
(c) The reagent $(CF_3CO_2)_2IPh$ converts $R—CONH_2 \rightarrow R—NH_2$.
Using this information, suggest how the tetrapeptide from (b) might be converted into B. (NB. Leucine amide, Leu—NH_2, is readily available, but $R—CH(NH_2)_2$ is unstable unless one or both nitrogens are acylated.)
(d) How would you confirm that your synthetic peptide B had the following features:

(i) That Tyr was the N-terminus?
(ii) That there was no free C-terminus?
(iii) That is contained the amino acid residues of Glu, Gly, Phe, and Tyr (and in equal amounts)?
(iv) That it was cyclic?

CHAPTER 7

The Structure of LH-RH

In 1977, the Nobel Prize for Medicine was shared by Roger Guillemin and Andrew Schally, for their work on the structure of **luteinising hormone releasing hormone** (LH-RH). The elucidation of this structure was the culmination of 15 years work for Guillemin and Schally. But the two men did not work together on the structure determination; indeed, it was their rivalry that spurred them on, with Schally being the first to arrive at a structure for pig LH-RH. It is the work of his group that we will look at in detail in this chapter.

1   What is LH-RH?

At the base of the brain in all mammals is a small gland called the **pituitary**. This gland releases a range of **hormones**, which control many features of growth and reproduction.

> **Hormones** are chemicals that are released in one part of the body, are carried through the circulatory system, and which influence cells elsewhere in the body. Absolutely minute quantities of hormones are often enough to elicit large cellular responses. They are used by the body to control the growth and development of most organs.

Immediately above the pituitary, there is another gland-like entity called the **hypothalamus**, whose function was unknown in the early 1950s.

By the late 1950s, the suggestion had been put forward that the hypothalamus was itself a gland, releasing hormones that controlled the pituitary. The idea was not only a very controversial one, but also one that was recognised to be very hard to prove. The hormones from the pituitary are hard to detect and isolate because they are produced in such small quantities—cells often only need a few **molecules** of a hormone in order to respond. By analogy, the amounts of hormones produced by the hypothalamus would be expected to be unbelievably small—how could they ever be isolated and studied?

There was one feature of hormones from the hypothalamus that could be utilised: they would presumably have a big effect on pituitary cells, causing them

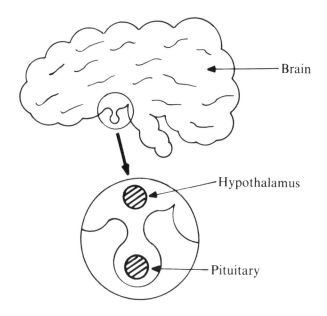

to release other hormones. By the 1960s, extremely sensitive bio-assays had been developed, that allowed the presence of very small amounts of known hormones to be detected (using antibody binding properties). Schally (and Guillemin) decided to look for a hormone produced by the hypothalamus, that would stimulate pituitary cells to release luteinising hormone—a polypeptide that controls the growth and functioning of testes or ovaries, and whose presence could be easily monitored.

> **LH-RH** is so called because it is the **Hormone** that causes the **Release** of **Luteinising Hormone**.

2   How was LH-RH Isolated?

Schally decided to try and isolate LH-RH from pigs, starting from hypothalami collected from abbatoirs. The isolation of pure LH-RH turned out to be by far the most time-consuming part of the research. And the amounts of LH-RH were so small that its presence could not be detected by UV absorption, reaction with ninhydrin, or any other normal chemical method. At every stage of purification, the hypothalamus extracts had to be fractionated, and each fraction tested for LH-RH activity using the bio-assay technique. In the end, Schally developed a 12-stage purification procedure, as outlined below:

Solvent extraction	Gel filtration	Ion exchange chromatography	Electrophoresis
1			
2			
	3		
4			
		5	
		6	
			7
8			
	9		
	10		
			11

Table 7.1.
After isolation and pulverisation of the hypothalami, this material (2.5 kg) was subjected to 11 further purification steps. Fractions were tested for LH-RH activity by measuring the amount of LH released by rats (using a radio-immuno assay). The final product (830 µg) had a peptide content of only 31%, indicating that only 260 µg of LH-RH had been obtained.

How much LH-RH did Schally obtain? In the end, he had to extract LH-RH from a total of 165,000 pigs. And he obtained..... 260 µg!

To get this in perspective, consider how much natually occurring human LH-RH there is in the world. If humans are about as sexually active as pigs, then we might assume that they produce about the same amount of LH-RH.

So 1 million humans $\longrightarrow$ 1.5 mg of LH-RH

And 4000 million people $\longrightarrow$ 6 g

Therefore, the entire human race only has a few grams of LH-RH. As an aphrodisiac, LH-RH must be quite potent!

Schally's problem, however, was how to determine the structure of LH-RH with only 260 µg of material.

3 How was the Structure of LH-RH Determined?

3.1 Amino Acid Analysis

By hydrolysing just 10 µg of LH-RH in concentrated HCl, Schally's group were able to carry out an amino acid analysis. They concluded that LH-RH was composed of nine amino acids:

Arg, Glu, Gly(2), His, Leu, Pro, Ser, Tyr

3.2 End Group Analysis

This completely failed! Reaction of LH-RH with dansyl chloride, followed by acidic hydrolysis, failed to give any dansyl derivatives; similarly, treatment with carboxypeptidase did not cause C-terminal amino acids to be liberated.

This meant that both ends of the peptide were blocked, so that structure determination would be extremely difficult. Indeed, it was not thought that $260\,\mu g$ would be enough material for the sequencing to be carried out—until Schally recruited the expertise of a Japanese chemist called Matsuo.

3.3 Fragmentation of LH-RH

In order to obtain at least one free N^{α}-amino group (needed for sequencing to be possible by Edman degradation), Matsuo decided to fragment LH-RH enzymically—each point of cleavage should generate a free N- and a free C-terminus which could be detected. But because the quantities of LH-RH were so small, Matsuo employed a new, highly sensitive radio-labelling method of identifying the C-termini:

(i) Step 1. Form the oxazolone (see pages 127–128)

156

(ii) Step 2. React oxazolone with tritium labelled H₂O

(iii) Step 3. Carry out amino acid analysis. The C-terminal α-position is tritium labelled, and this is retained during hydrolysis. The only radioactive amino acid is therefore derived from the C-terminal residue.

In conjunction with the dansyl method of identifying N-termini, Matsuo was able to digest small amounts (5 μg samples) of LH-RH enzymically, and analyse for the N- and C-termini.

Using **trypsin** (which normally cleaves on the carboxy side of basic residues), no fragmentation was observed. But when **chymotrypsin** was employed, an extraordinary result was obtained: **two N^α-dansyl derivatives** were identified, but only **one radio-labelled C-terminal amino acid** was observed. This 'impossible' result was finally solved by Matsuo, who realised that chymotrypsin cleaves peptides on the carboxy side of aromatic residues—in particular, phenylalanine, tyrosine, and **tryptophan**. Tryptophan is destroyed during acidic hydrolysis (i.e. by the conditions used before amino acid analysis), and nobody had checked for its presence. By carrying out alkaline hydrolysis instead, Matsuo showed that LH-RH contained 10 amino acids, with tryptophan being the missing residue. Amazingly, Guillemin's group made exactly the same mistake, but failed to realise it until much later.

Another enzyme, thermolysin, was found to cleave LH-RH in three places.

With the amino acid composition determined, and the knowledge that LH-RH could be specifically fragmented using chymotrypsin or thermolysin, Schally's group were in a strong position to start sequencing LH-RH.

3.4 Sequencing LH-RH

Even though LH-RH could be cleaved into fragments with free N-termini, two serious problems still remained:

(i) The Edman degradation was not sufficiently sensitive to detect the tiny amounts of thiohydantoins that would be produced.

(ii) There was so little LH-RH available that it was unreasonable to suppose that the fragments could actually be separated and purified.

In order to overcome the first problem, a combination of the Edman degradation and the (more sensitive) dansyl end group analysis was employed:

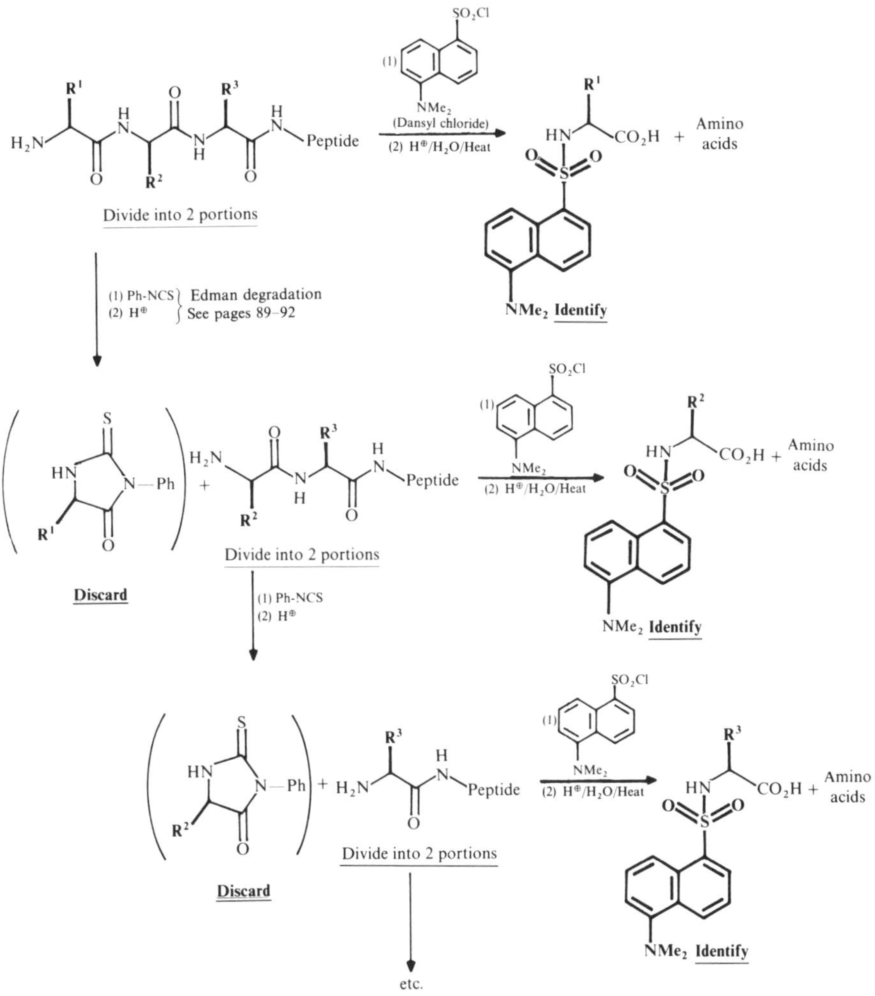

Concerning the separation of the fragments, an enormous gamble was taken. Not only was the analysis carried out on the **mixture** of fragments, but the **enzymic hydrolysis was stopped before completion**; although this gave quite a complex mixture of peptides, it in fact simplified the sequencing problem. To see how this came about, we can look at the results from their first successful partial hydrolysis, using chymotrypsin.

Chymotrypsin cleaves LH-RH in two positions. If the hydrolysis is stopped before completion, then a total of six peptides would be expected:

LH-RH decapeptide (A → J), with N- and C-termini blocked (|)

$$\vdash A - B - C - D - E - F - G - H - I - J - \vert$$

Chymotrypsin

Partial enzymic digestion

$$\vdash A - B - C - D - E - F - G - H - I - J - \vert$$
F1

$$\vdash A - B - C \qquad D - E \qquad F - G - H - I - J - \vert$$
F2   F3   F4

$$\vdash A - B - C - D - E$$
F5

$$D - E - F - G - H - I - J - \vert$$
F6

However, for sequencing studies from the N-terminus, three of these peptides (F1, F2 and F5) would fail to react (N-terminus still blocked); and two of them (F3 and F6) would produce an identical sequence of residues (until F3 runs out). So, as far as N-terminal sequencing is concerned, the mixture would react as if it were composed only of peptides F4 and F6. The C-terminal end group analysis would identify the residues immediately **before** fragments F4 and F6 (i.e. C and E).

The only free C-termini

→ E   F — G — H — I — J — |
F4

→ C   D — E — F — G — H — I — J — |
F6

Apparent composition of fragments for N-terminal sequencing after partial chymotryptic digestion of LH-RH.

But fragments F4 and F6 have overlapping sequences. If Schally's group were lucky, they hoped that some of the amino acids in their initial sequencing would reappear in the later analyses, and that these overlapping regions would allow them to determine part of the sequence of LH-RH in an unambiguous way. These were the results they obtained:

Table 7.2. Sequencing results on LH-RH after partial chymotryptic digestion.

	Termini	Detection method	Residues
	C-Termini	³H Labelling	Trp, **Tyr**
	N-Termini	(Dansyl chloride)	**Gly**, Ser
After 1st Edman degradation	N-Termini	(Dansyl chloride)	Leu, **Tyr**
After 2nd Edman degradation	N-Termini	(Dansyl chloride)	Arg, **Gly**
After 3rd Edman degradation	N-Termini	(Dansyl chloride)	Below detection limit

As you can see, the C-terminal tyrosine reappeared as an N-terminus after a single Edman degradation. So tyrosine must have been immediately before the first residue of F4, and the two cleavage sites must have been just two residues apart. It is also easy to see the Tyr—Gly sequence appearing twice; so we can start to fill in the sequences of F4 and F6:

Free C-termini {
→ Tyr   Gly—G—H—I—J— | F4
→ C   D—**Tyr**—**Gly**—G—H—I—J— | F6
} Fragments for N-terminal sequencing

By a process of elimination, peptide F6 must be preceded by Trp, and must start with Ser. We can now sequence F6 as Ser—Tyr—Gly, so that F4 must be Gly—Leu—Arg.

Free C-termini {
→ Tyr   Gly—Leu—Arg—I—J— | F4
→ Trp   Ser—**Tyr**—**Gly**—G—H—I—J— | F6
} Fragments for N-terminal sequencing

Using this logic, Schally's group concluded that LH-RH must contain the sequence:

—Trp—Ser—Tyr—Gly—Leu—Arg—

This was as much information as they could obtain from the hydrolysis by chymotrypsin.

So, turning to the enzyme thermolysin, they again carried out a partial hydrolysis. The results are shown below, and are then rearranged to show how they overlap with the results from the chymotrypsin experiment:

Table 7.3. Sequencing results on LH-RH after partial digestion with thermolysin.

	Termini	Residues	cf. Chymotrypsin
	C-Termini	Gly, His, Ser	Gly Ser His
	N-Termini	Leu, Trp, Tyr	Leu Trp Tyr
After 1 Edman	N-Termini	Arg, Gly, Ser	Arg Ser Gly
After 2 Edmans	N-Termini	Leu, Pro, Tyr	Tyr Leu Pro
After 3 Edmans	N-Termini	Only Gly detectable	

You should not find it hard to show that the sequencing of LH-RH could be extended to eight residues:

$$—His—Trp—Ser—Tyr—Gly—Leu—Arg—Pro—$$

Only the positions of a glutamic acid and a glycine residue still needed to be determined, and the nature of the blocking groups. At this stage, Schally's group had exhausted the information they could gain from hydrolysis/sequencing experiments.

3.5 Other Results

The amounts of LH-RH were too small for spectroscopic studies, but mass spectrometry yielded one crucial clue: there were strong peaks at m/z 112 and 84. Only one residue gives these characteristic signals; pyroglutamic acid (Glp), which must have been at the N-terminus (see page 108).

This showed that the Glu from the amino acid analysis occupied the N-terminal position in LH-RH, and that the N^{α}-amino group was blocked via the cyclic amide (see page 161).

Schally's group guessed that the C-terminus was probably blocked as the amide, and this left only two possible structures:

$$LH\text{-}RH \begin{cases} Glp—\textbf{Gly}—His—Trp—Ser—Tyr—Gly—Leu—Arg—Pro—NH_2 \\ \qquad\qquad\qquad\qquad OR \\ Glp—His—Trp—Ser—Tyr—Gly—Leu—Arg—Pro—\textbf{Gly}—NH_2 \end{cases}$$

Their analytical methods had led them almost to the answer, but they could not differentiate between these two structures. The only solution was to synthesise both peptides, and see which one possessed the desired biological properties.

4   The Synthesis of LH-RH

Schally's group wanted to synthesise rapidly the two possible candidates for LH-RH, in order to be the first to announce the structure. And because they needed only minute amounts of the peptides in order to detect biological activity, they used the solid-phase technique, employing the standard Merrifield approach (see Figure 7.1 on page 162).

Within two weeks, they had prepared a synthetic peptide with identical biological properties to LH-RH—the structure was solved! It turned out that the glycine was at the C-terminus, so that the final structure was:

Glp—His—Trp—Ser—Tyr—Gly—Leu—Arg—Pro—Gly—NH$_2$

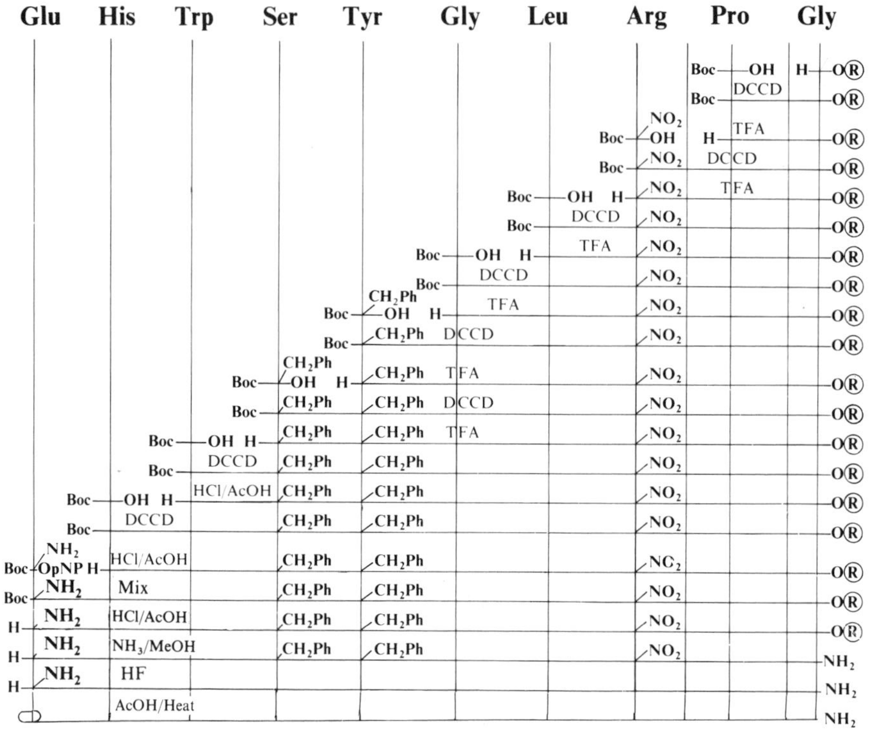

Glu	His	Trp	Ser	Tyr	Gly	Leu	Arg	Pro	Gly
								Boc——OH	H—O(R)
								Boc DCCD	O(R)
							NO₂ Boc—OH	H TFA	O(R)
							Boc /NO₂	DCCD	O(R)
						Boc——OH H	/NO₂	TFA	O(R)
						Boc DCCD	/NO₂		O(R)
					Boc——OH H	TFA	/NO₂		O(R)
				Boc—OH Boc CH₂Ph	DCCD		/NO₂		O(R)
				Boc /CH₂Ph—OH H	/CH₂Ph TFA		/NO₂		O(R)
				Boc /CH₂Ph	/CH₂Ph DCCD		/NO₂		O(R)
			Boc CH₂Ph /OH H	Boc /CH₂Ph—OH H	/CH₂Ph TFA		/NO₂		O(R)
			Boc /CH₂Ph	/CH₂Ph	/CH₂Ph DCCD		/NO₂		O(R)
		Boc——OH H	/CH₂Ph	/CH₂Ph	/CH₂Ph TFA		/NO₂		O(R)
		Boc DCCD	/CH₂Ph	/CH₂Ph	/CH₂Ph		/NO₂		O(R)
Boc——OH H	HCl/AcOH	/CH₂Ph	/CH₂Ph	/CH₂Ph		/NO₂		O(R)
Boc DCCD		/CH₂Ph	/CH₂Ph	/CH₂Ph		/NO₂		O(R)
Boc /NH₂ OpNP H	HCl/AcOH	/CH₂Ph	/CH₂Ph	/CH₂Ph		/NO₂		O(R)
Boc /NH₂	Mix	/CH₂Ph	/CH₂Ph	/CH₂Ph		/NO₂		O(R)
H /NH₂	HCl/AcOH	/CH₂Ph	/CH₂Ph	/CH₂Ph		/NO₂		O(R)
H /NH₂	NH₃/MeOH	/CH₂Ph	/CH₂Ph	/CH₂Ph		/NO₂		O(R)
H /NH₂	HF								NH₂
	AcOH/Heat								NH₂
									NH₂

Figure 7.1. Solid-phase peptide synthesis of LH-RH.

Incredibly, they had used only 50 μg of the LH-RH isolated from natural sources.

When Schally's group announced the structure of pig LH-RH, Guillemin's group were still unaware that tryptophan was present in their peptide. With that vital information, they were able to solve the structure of their sheep LH-RH within 2 months, and showed that it was identical to pig LH-RH.

5 Medicinal Properties of LH-RH

Although the discovery and structure elucidation of LH-RH was a major breakthrough in peptide chemistry, the medicinal value of LH-RH was not immediately apparent. The known biological properties of LH-RH (due to its luteinising hormone releasing effect on the pituitary) were:

(i) The control of the growth and development of ovaries or testes.
(ii) The triggering of the release of other hormones—in particular, oestrogen and testosterone.

Around 1980, it was found that certain cancers responded dramatically to levels of specific hormones. In particular, prostate cancer could be controlled by lowering the levels of testosterone. It was therefore proposed that an analogue of LH-RH might be developed, which would bind to the active site of LH-RH receptors, but would not trigger the release of testosterone; compounds of this type (known as **antagonists**) would block the action of naturally occurring LH-RH, and thereby reduce the levels of testosterone produced.

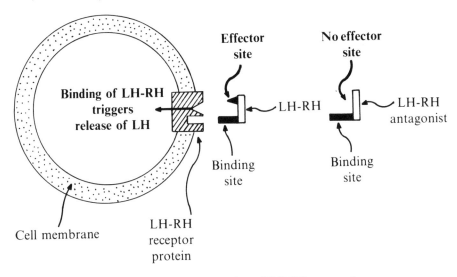

Figure 7.2. Schematic drawing of LH-RH antagonist

A simple LH-RH antagonist was not discovered, but it was found that some analogues were considerably more potent than LH-RH itself. Interestingly, after an initial increase in testosterone levels, these **super-agonists** caused the hormone release mechanism to turn itself off (a process known as **down-regulation**). By administering a steady level of these highly active LH-RH analogues, it proved possible to dramatically suppress the levels of testosterone.

One of the super-agonist LH-RH analogues, Zoladex^R, has been through full clinical trials, and is now available for the treatment of prostate cancer.

Glp—His—Trp—Ser—Tyr—**D-Ser(Bu^t)**—Leu—Arg—Pro —**AzGly**—NH$_2$

Zoladex is a 'super-agonist' analogue of LH-RH, in which the two glycine residues have been replaced by unusual amino acids that make the peptide less susceptible to hydrolytic breakdown.

Some pharmacological terms.

Agonist. A compound that triggers the biological response being studied.

Antagonist. A compound that blocks the receptor (without triggering biological response); such molecules inhibit the effect of agonists.

Maximal response. How ever much agonist is administered, there is always a maximum possible biological response (usually when all the receptor sites are filled by agonist molecules)—this is the maximal response.

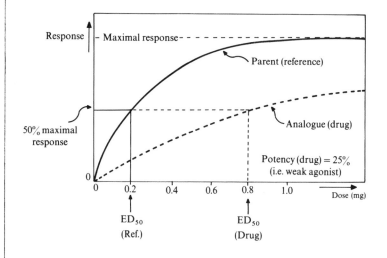

ED_{50}. This is the dose that causes 50% of the maximal response.

Affinity. This is a quantitative measure of the equilibrium constant for a drug/receptor (DR) complex:

$$D + R \underset{k_{-1}}{\overset{k_1}{\rightleftharpoons}} DR$$

$$\text{affinity} = K = \frac{[DR]}{[D][R]}$$

At 50% response, $[DR] = [R]$ (ideally)

$$\therefore \quad K = \frac{1}{[D]} = \frac{1}{ED_{50}}$$

Potency. This is the relative affinity of a drug for a specific receptor, compared with a reference compound (often the 'parent' natural product); i.e.

$$\text{Potency} = \frac{ED_{50}(\text{Ref.})}{ED_{50}(\text{Drug})}$$

So if, in order to get a 50% response (ED_{50}) from a drug, you need to administer four times as much as for the reference compound, then the drug's potency would be 0.25 (or 25%).

Plots of log (dose) vs. % (response) are often used; the resulting graphs are almost linear across a wide range of doses, so that ED_{50} can be determined more accurately. However, biochemical systems are usually much more complicated than the above definitions might imply; drugs sometimes bind to several different receptors, are rapidly degraded, or cause the observed response by an unexpected mechanism. So the full pharmacological assessment of a new drug is an enormous undertaking, and a great many factors must be considered and analysed.

Although many peptides have useful biological properties, there are many problems with their medicinal use. For example:

(i) Peptides can be degraded rapidly by peptidases in the body—in order to reduce this problem with Zoladex[R], the serine residue is a D-amino acid. A further problem is that they are often excreted quite rapidly through the kidneys. Overall, this means that it is hard to obtain long-term effects by administering a single dose of a peptidic drug.

(ii) Peptides cannot usually be given orally, due to poor absorption (and the presence of peptidases) in the gut. Zoladex[R] is actually administered via a pellet the size of a grain of rice, which is inserted under the skin; the pellet consists of a biodegradable polymer in which the Zoladex[R] is uniformly dispersed, and as the pellet gradually dissolves (over a period of a month or so), the 4 mg of Zoladex[R] is slowly released into the blood stream.

(iii) Side effects can also be a problem with peptides (as with any drug). However, by using solid-phase peptide synthesis, it is usually possible to prepare a large range of analogues; the structures and biological properties can then be correlated (**structure/activity relationships**—see Appendix B), guiding medicinal chemists in the design of drugs that are both selective and effective.

Further Reading

Nicholas Wade, *New Scientist*, 1981, **92**, 251. A good yarn! Tells the LH-RH story in an enjoyable and readable way.

A. V. Schally *et al.*, *Biochem. Biophys. Res. Commun.*, 1971, **43**, 393; *ibid.*, 1971, **43**, 1334; *ibid.*, 1971, **44**, 459; *ibid.*, 1971, **45**, 822. These four papers, all from the same journal, describe the isolation, structure elucidation, and synthesis of LH-RH by Schally's group.

A. S. Dutta *et al.*, *J. Med. Chem.*, 1978, **21**, 1018. Reports the synthesis and biological activity of Zoladex[R].

A. S. Dutta, *Drugs of the Future*, 1988, **13**, 43. Discusses how Zoladex[R] was developed into a usable drug.

A. S. Dutta, *Chem. Br.*, 1989, **25**, 159. A nice overview of 'Small peptides—New targets for drug research'.

APPENDIX A

Synthesis of Amino Acids

Whenever the chemical synthesis of a peptide is undertaken, it is necessary to have a source of the constituent amino acids. For simple DNA encoded amino acids, they are normally available from natural sources—although special routes have been developed to some of them. But many biologically active peptides contain unusual amino acids, either because they are present in the natural sequence (e.g. many antibiotics), or because they need to be introduced in order to produce improved medicinal properties. How are such amino acids synthesised?

In fact, it would be impossible to discuss all of the possible methods by which amino acids might be synthesised. In any case, the specific target molecule must be considered carefully, because special methods might be particularly suitable for it. Nevertheless, there are a few tactics that are used extremely widely; we will consider these first of all (with phenylalanine as a specific example where appropriate). Then we will look at the methods used in the preparation of optically active amino acids, using a range of specific examples to indicate the versatility of the techniques. Finally, we will look at one specific optically active amino acid that has been synthesised by several different methods in recent years.

1 General Methods of Amino Acid Synthesis

1.1 Strecker Synthesis

This is the classical way of preparing α-amino acids, and was developed by Strecker in the 1930s. The method involves the conversion of an aldehyde into the corresponding amino nitrile, which can then be hydrolysed to the amino acid (see next page). The reaction mixture needs to be buffered (usually with ammonium chloride), and is believed to proceed via an imine intermediate.

The Strecker synthesis is a very reliable way of converting aldehydes into the corresponding amino acids, and many variations of the method have been devised. One limitation, however, is that the hydrolysis of nitriles to carboxylic acids requires quite vigorous conditions (e.g. 6 M HCl, 100 °C, 10 h), so that sensitive side-chains might be destroyed.

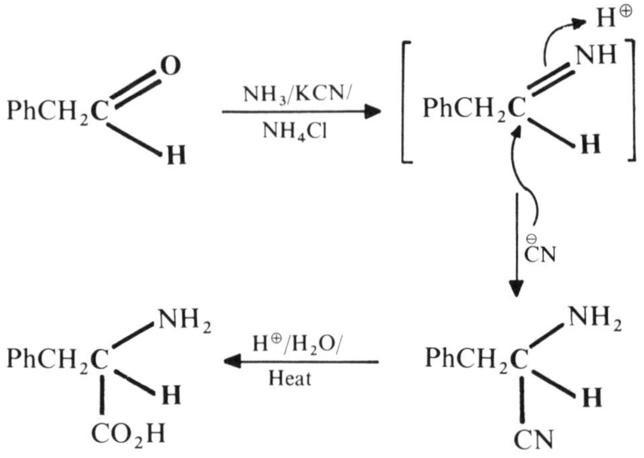

1.2 From Glycine α-Anion Equivalents

The other general route to α-amino acids involves starting from a derivative of glycine, removing the α-proton with base, and then carrying out a simple alkylation reaction:

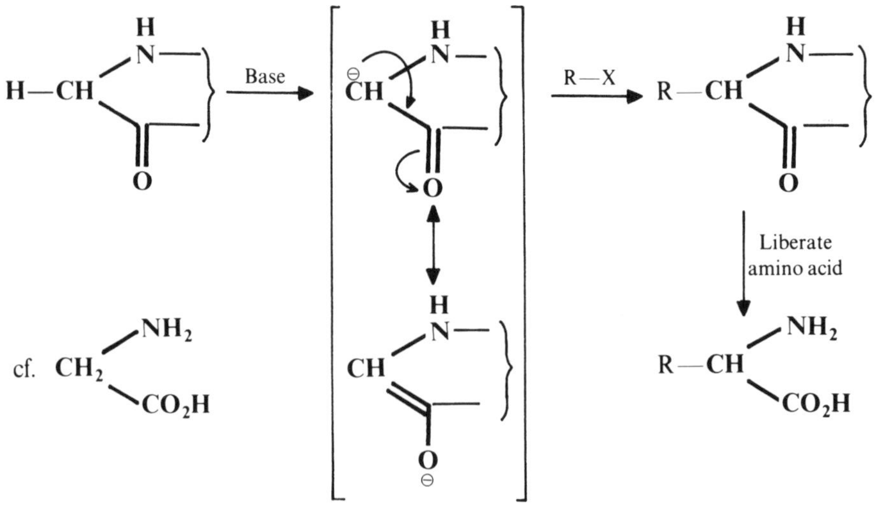

There are many examples of this tactic being utilised, the problem usually being to ensure that the amino nitrogen does not act as the nucleophile during the alkylation step. Three methods of generating glycine α-anion equivalents are given below, although many others exist.

1.2.1 *Oxazolones.* These compounds are usually a problem during coupling reactions in peptide chemistry, because the lability of the α-proton can lead to racemisation (see pages 127–28). This property can be used to advantage in the synthesis of amino acids, as exemplified below:

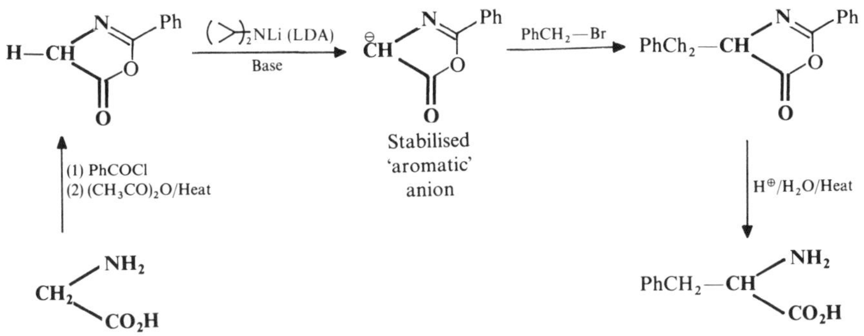

1.2.2 *α-Amidomalonic Ester.* Esters of malonic acid (1,3-propanedioic acid) can be readily deprotonated (e.g. with NaOMe) because the resulting anion is stabilised by two carbonyl groups; such anions can be readily alkylated.

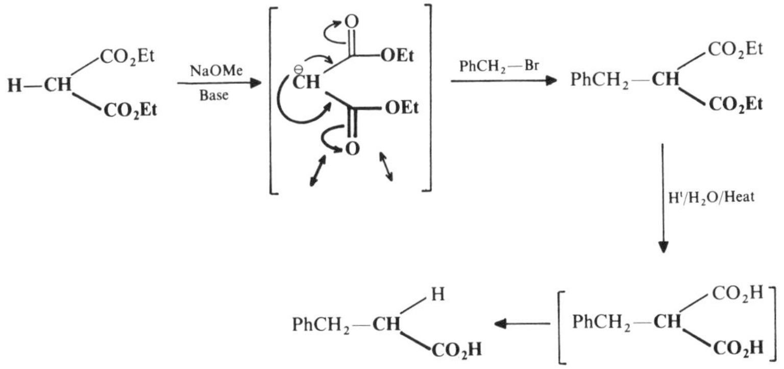

Hydrolysis of the resulting di-ester gives the di-acid, which (being a β-keto acid) readily loses CO_2 on heating to yield an α-alkylated ethanoic acid.

If this sequence of reactions is carried out using a protected α-amino derivative of malonic ester, then the final product is an α-alkylated glycine. For example, α-acetamidomalonic ester can be readily prepared from diethyl malonate via the following two-step procedure.

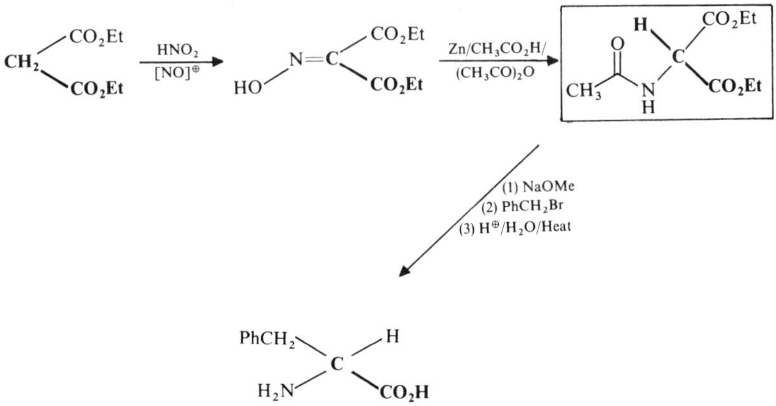

The most labile proton of α-amidomalonic ester is the α-proton (rather than the NH-proton). So treatment of the amino di-ester with the alkylation/decarboxylation procedure gives the corresponding amino acid. Because α-amidomalonic ester is so readily prepared (or bought), this method is quite widely used.

1.2.3 Fully Protected Glycine.

If the amino group of glycine is significantly nucleophilic (e.g. primary or secondary amine), or can be readily deprotonated (e.g. simple amides), then α-alkylation cannot be carried out (N-alkylation occurring instead). However, N,N-dibenzyl derivatives of glycine are virtually non-nucleophilic (tertiary amine), have no NH-protons, and can be debenzylated by catalytic hydrogenolysis. It is therefore an excellent glycine derivative for the synthesis of amino acids:

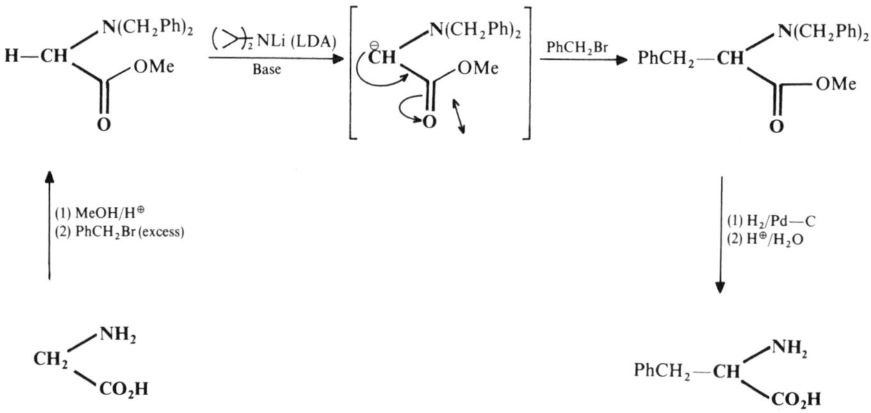

For example, after formation of the dibenzyl derivative of simple glycine esters, α-deprotonation can be effected with a powerful base (e.g. lithium diisopropylamide—LDA). After alkylation of the anion, removal of the dibenzyl protection yields α-alkylated glycine derivatives.

1.3 Other Methods

These are virtually too numerous to mention, with many of them highly specific to particular target molecules. But a number of other tactics have been used quite extensively, and a small selection of them are shown below.

1.3.1 *Via the Hydration*

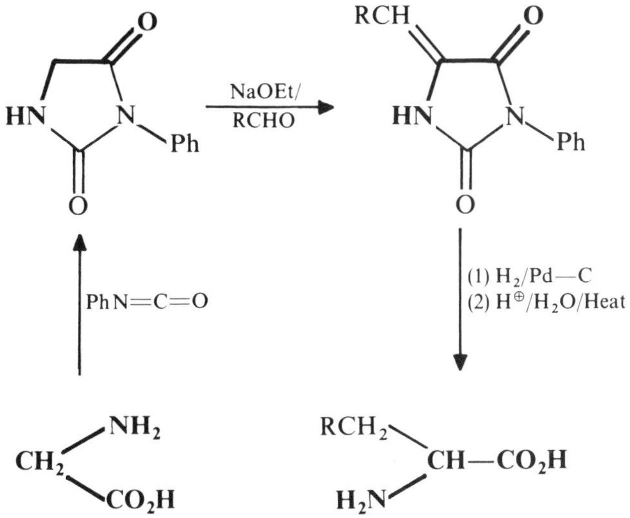

1.3.2 *Using α-Keto Acids*

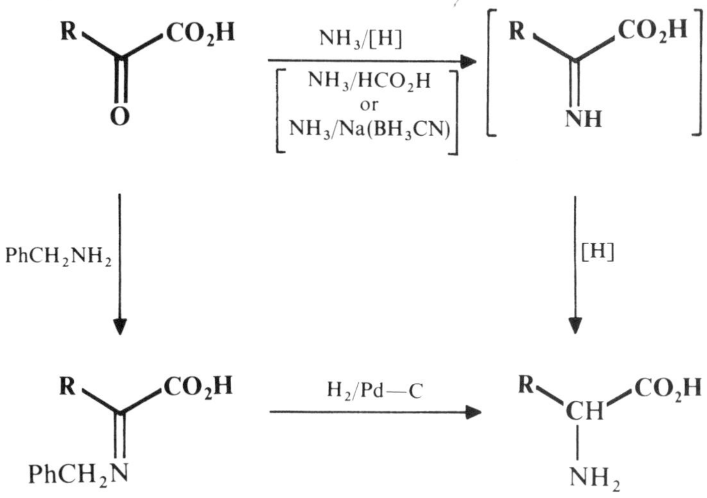

1.3.3 *Via α-Bromo Esters*

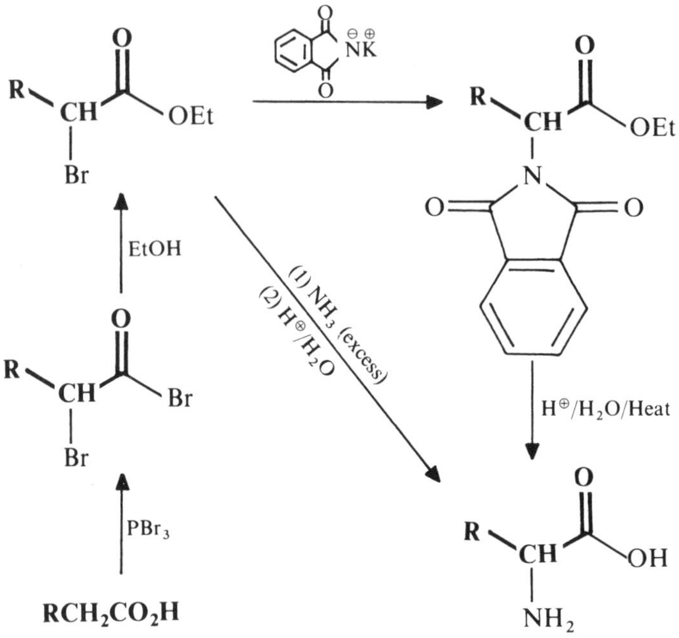

2 Preparation of Optically Active Amino Acids

Almost all of the naturally occurring amino acids are optically active, with the obvious exception of glycine. The structure and biological properties of peptides depend crucially on the stereochemistry of each of the residues. So, in order to fully explore the biological properties of new peptides, it is important that any chiral components should be optically pure (if possible).

Optical purity. This is usually expressed as the **enantiomeric excess** (e.e.)—the percentage by which one of the mirror images is in excess over the other. If a compound is composed of $A\%$ of the major enantiomer, and $B\%$ of the minor enantiomer, then:

$$\text{e.e.} = (A - B)\%$$

The synthetic strategies outlined in the previous section lead to racemic amino acids, and special tactics are necessary in order to generate optically active compounds; there are three approaches.

2.1 Resolution of Racemic Mixtures

The simplest synthetic tactic is to prepare the target molecule in racemic form, and then to separate the enantiomers afterwards. This **resolution** step can be carried out in one of a number of ways.

2.1.1 Enzymic Resolution. This is widely used for amino acids, because there are several readily available enzymes that selectively cleave the amide bond of L-amino acids. The usual approach is to form the N-ethanoyl (acetyl) derivative of a D, L-mixture of amino acids, and then to treat with an acylase. Only the L-acyl derivative is hydrolysed, leaving the D-residue still protected.

Enzymic resolution of N^α-acetyl amino acids.

The two products can be readily separated; the free amino acid is soluble in aqueous acid, whilst the ethanoyl derivative can be extracted into an organic phase.

If both the D- and L-amino acids are required, then the ethanoyl group can be hydrolysed from the D-enantiomer. If only one optical isomer is needed, then the unwanted enantiomer can be racemised (e.g. see pages 126–129) in order to produce more racemic material for resolution.

2.1.2 Crystallisation with a Chiral Counter-ion. Optical isomers cannot be separated without using a chiral compound (e.g. an enzyme) which can interact **differentially** with the two enantiomers. For amino acids, it is sometimes possible to form a salt with a chiral counter-ion which preferentially crystallises out with the D- or L-amino acid. These salts can be formed with chiral acids or bases, which are themselves usually derived from natural sources.

Crystallisation of racemic amino acids with chiral counter-ions. For neutral amino acids, the N^α-Z-protected derivatives are often separable as the (R)-1-phenylethylammonium salts (A), whilst basic amino acids can sometimes be resolved directly using (1S)-10-camphorsulphonic acid (**B**).

Successful resolution by this method often requires repeated **fractional crystallisation.** There are no rules about which enantiomer is more likely to crystallise out, nor about the conditions necessary for crystallisation to occur. So the method requires considerable trial and error. But if successful, it does allow large quantities of enantiomers to be separated, and this has ensured that this method has remained an important one.

2.1.3 *Chiral HPLC.* Normal HPLC is, of course, unable to distinguish between enantiomers. But if a chiral group is bound to the stationary phase, then resolution of optical isomers becomes possible. Several chiral HPLC columns are now available, although they are mainly used for analytical work (e.g. calculating e.e.s) rather than for preparative work. Columns using carbohydrates or proteins as the chiral components of the stationary phase are available, but ones using amino acid derivatives linked to a silica support are more widely used:

**The Pirkle chiral stationary phase, for
resolution of enantiomers using HPLC.**

The resolution of the enantiomers of protected amino acids is often excellent
using these columns.

2.1.4 *Separation of Diastereotopic Peptides.* If only one of the constituent amino
acids in a peptide is racemic, then it is often possible to remove the unwanted
isomer by standard purification methods (e.g. chromatography) **after** formation
of the peptide. This is because the other chiral centres will interact differently with
D- and L-residues, leading to **diastereoisomers** with different physical properties.
This tactic is extremely simple to carry out, and the separation of diastereoisomers can be attempted at several stages in the synthesis of the peptide—
but it is nevertheless quite a gamble, and resolution of the racemic amino acid is
usually attempted first.

2.2 Asymmetric Synthesis

Asymmetric synthesis is the formation of a chiral compound in which one of the
enantiomers predominates, and it is an area of organic chemistry that has
received considerable attention in recent years. In contrast to standard synthesis
followed by resolution, asymmetric synthesis is the direct generation of an
optically target molecule at the end of a synthetic sequence. This can be achieved
by one of two methods.

2.2.1 *Using Chiral Reagents.* A wide range of chiral reagents have been
developed; e.g. oxidising agents, reducing agents, and catalysts (including
enzymes). For example, hydrogenation of a didehydro derivative of phenylalanine can yield the protected amino acid in 96% e.e.

(R, R)-Dipamp

(*J. Chem. Soc., Chem. Commun.*, 1972, 10)

The use of specific chiral reagents depends upon the particular target molecule. For example, some β-amino acid derivatives are accessible via the Sharpless asymmetric epoxidation—a reaction that generates chiral epoxides with high e.e. from allyl alcohols:

Above is shown the principle of the asymmetric synthesis of (2R, 3S)-3-amino-2-hydroxy-3-phenylpropanoic acid; the details of the actual sequence of reactions are given below:

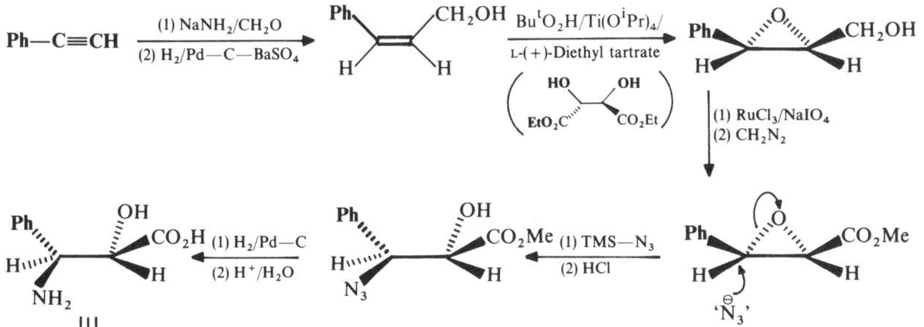

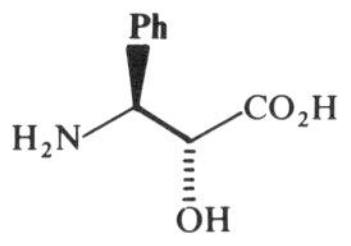

One diastereoisomer
Yield ≃ 25% (from Ph—C≡CH)
Enantiomeric excess > 95%

(*J. Org. Chem.*, 1986, **51**, 46)

2.2.2 *Using Chiral Auxiliaries.* This tactic employs a removable chiral group which is attached to the starting material; it directs the stereochemistry during the formation of the new chiral centre, leading to a predominance of one of the stereoisomers. A wide range of chiral auxiliaries have been developed, of which two examples have been selected:

Example 1

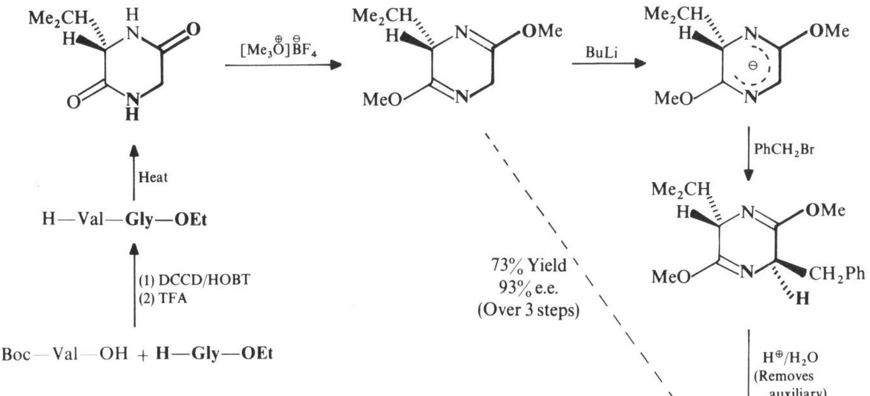

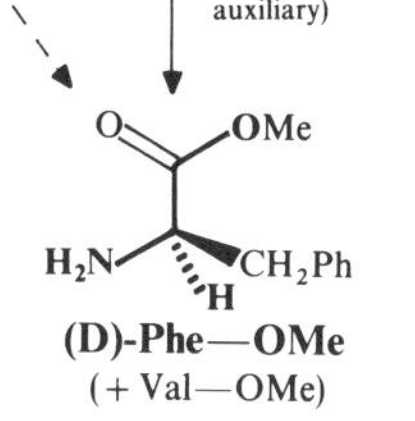

73% Yield
93% e.e.
(Over 3 steps)

(U. Schöllkopf, *Pure Appl. Chem.*, 1983, **55**, 1799)

(D)-Phe—OMe
(+ Val—OMe)

Example 2

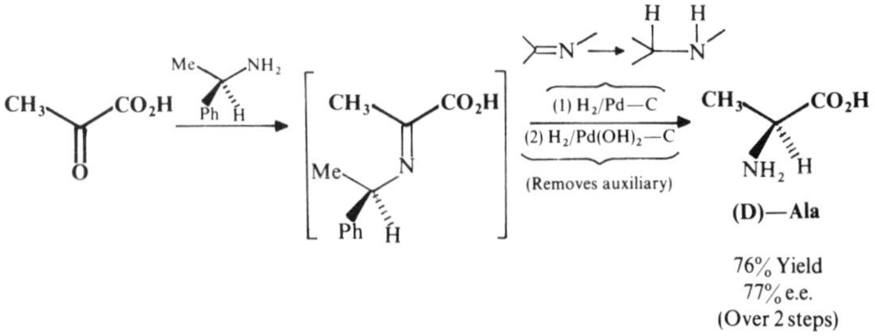

76% Yield
77% e.e.
(Over 2 steps)

(For asymmetric hydrogenations, see
Asymmetric Synthesis, Volume 5 (*Chiral Catalysis*),
J.D. Morrison (Ed.), Chapter 3 (pp. 71—101) and Chapter 10
(pp. 345–383), Academic Press, 1985)

After the new chiral centre has been formed, the chiral auxiliary can be removed. But this step is not usually carried out until after the compound has been purified—the chirality of the auxiliary ensures that any mixture of stereochemistry at the new chiral centre generates diastereoisomers, which should have different physical properties. So it is often possible to use a chiral auxiliary both to induce asymmetry at the new chiral centre, and to simplify purification to a single stereoisomer.

2.3 Chiral Synthesis

This approach is a special form of asymmetric synthesis, in which the starting material is itself optically active. Its chiral centre is then incorporated into the target molecule. An obvious starting point for many unusual amino acids is an optically pure, readily available amino acid (such as those encoded by DNA). As with all the examples in this appendix, the viability of this approach depends crucially upon the specific target molecule, as exemplified below.

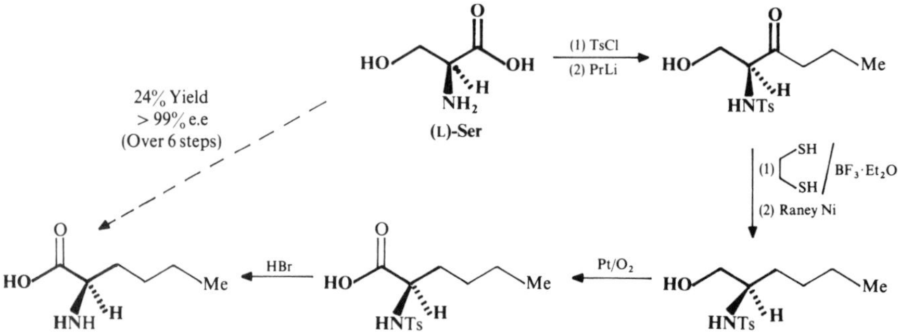

(D)-2-Aminohexanoic acid

(*J. Amer. Chem. Soc.*, 1984, **106**, 1095)

So, for any unusual amino acid that requires synthesis, there is an enormous range of synthetic approaches available, and many tactics for generating optically active products. But the specific route chosen must be devised by careful consideration of the actual structure being tackled, and with reference to previous literature on syntheses of related compounds. For the final example shown below, many different tactics were viable, all with certain advantages and disadvantages.

3 Synthesis of GABOB

γ-Aminobutyric acid (GABA) is a neurotransmitter, involved in the modulation of nerve impulses:

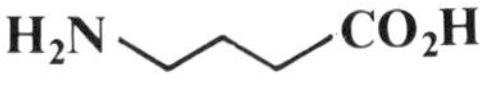

$$H_2N \diagdown \diagup \diagdown \diagup CO_2H$$

GABA

Many derivatives of GABA have been studied in recent years, because of their medicinal potential. One of these is the 3-hydroxy derivative of GABA known as GABOB. GABOB is also a neurotransmitter, but the (R)-enantiomer shows a much higher biological activity than the S-isomer, and pharmacologists wished to study the properties of pure (R)-GABOB in more detail.

Initially, GABOB was synthesised in racemic form using standard transformations. Then, in the late 1970s, resolution of racemic GABOB was achieved by crystallisation of the camphorsulphonic salts of the amide derivative.

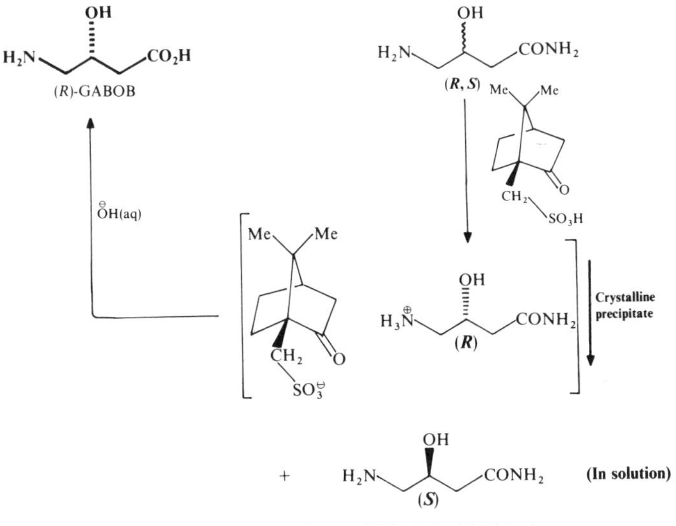

(Japanese patent, *Chem. Abs.*, 1977, **86**, 89207u)

This gave access to optically pure (R)-GABOB, but it was nevertheless a wasteful and time-consuming procedure.

In 1980, a chiral synthesis from ascorbic acid (vitamin C) was published, but it was long (nine steps) and low yielding (ca. 10%). But in 1983, a three-step asymmetric route was developed, using asymmetric epoxidation as the key step.

17% Yield
55% e.e.
(Over 3 steps)

(B.E. Rossiter, PhD Thesis, Stanford, 1981)

Despite being short, this route was low yielding, and gave only a moderate enantiomeric excess. So the three-step enzymic sequence published in 1985 was an obvious improvement.

45% Yield
> 95% e.e.
(Over 3 steps)

(*J. Amer. Chem. Soc.*, 1983, **105**, 5925)

It took much effort to determine the best β-keto ester for reduction by baker's yeast, but a very high e.e. was eventually achieved. The synthetic target was in fact the trimethylammonium derivative of (R)-GABOB, but the approach was clearly applicable to (R)-GABOB itself too.

The best e.e. achieved to date involves chiral synthesis from malic acid. Both R- and (S)-malic acid are readily available, although this synthesis (published in 1987) used the cheaper (S)-isomer to yield (S)-GABOB:

25% Yield
≃ 100% e.e.
(Over 8 steps)

(*J. Org. Chem.*, 1985, **50**, 5480)

The various routes to (R)-GABOB show the range of methods that are available for the asymmetric synthesis of amino acids. Sometimes optical purity is of crucial importance (if the wrong enantiomer is toxic, for example, as with thalidomide); sometimes yield is more important, or the versatility of the method for preparing new analogues. But the range of techniques that are now available mean that a satisfactory synthetic route to most amino acids is usually viable.

Further Reading

The syntheses of specific amino acids in the Appendix have been referenced under the appropriate scheme. More general coverage includes:
Comprehensive Organic Chemistry, Volume 2, pp. 815–40 (I.O. Sutherland, Ed.) and Volume 5, pp. 187–214 (E. Haslam, Ed.), Pergamon Press, 1979.
Chemistry and Biochemistry of the Amino Acids (G.C. Barrett, Ed.), Chapter 8, 'Synthesis of amino acids', pp. 246–96, and Chapter 10, 'Resolution of amino acids', pp. 338–53, Chapman and Hall, 1985.
Synthesis of Optically Active α-Amino Acids, R.M. Williams, Pergamon Press, 1989.

APPENDIX B

The Structure of Peptides

Proteins generally have well-defined three-dimensional structures, which are crucial for their biological activity. With peptides, on the other hand, the molecular structure usually has considerable freedom of movement, and tertiary structure is rarely observed. That is not to say that the **stereochemistry** and **conformation** of peptides are not important—on the contrary, peptides are usually extremely sensitive to such features. But their **active** conformation is often only achieved when they are actually bound to the target molecule, which is usually a protein—often an enzyme or a receptor on the outside of a cell.

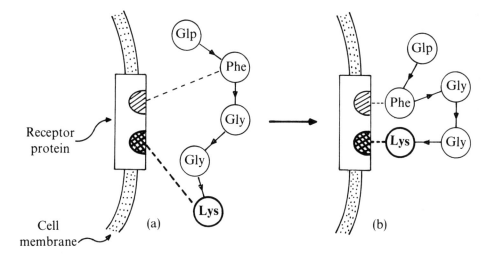

Schematic representation of a randomly coiled pentapeptide (a) adopting a well defined conformation (b) [due to binding with hydrophobic (⊘) and anionic (✪) sites on a receptor protein].

We can divide our consideration of the stereochemistry of peptides into three aspects: firstly, the stereochemical features that are typical of peptides in general; secondly, key features of secondary structure that are common among peptides; thirdly, methods of determining structural features of peptides.

1 General Stereochemical Features of Peptides

The first important feature is that peptide bonds are **planar**. This is due to resonance stabilisation of the amide bond (see pages 4–5). Moreover, the amide bond is usually *trans* (except with proline residues or highly strained cyclic peptides, when *cis* amide bonds sometimes occur).

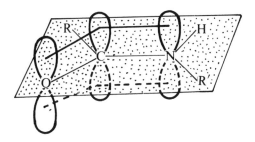

Planar amide bond. All the σ bonds lie in one plane; the π-bonds are above and below the plane. The R and R' groups prefer to be *trans*.

The first part of peptide structure that is usually considered is the backbone, i.e. only the amide bonds and the α-carbons, but not the side-chains:

The shape of this backbone can be completely defined by simply stating the torsional angles at each of the positions 1–8 above (if we assume that all the peptide bonds are *trans* and *planar*).

Newman projection

Torsional angle (θ) between bonds a and c in a conformer of 1-chloropropane. When **Cl** and **CH$_3$** are eclipsed, $\theta = 0°$.

> **Torsional angle.** If we label three consecutive bonds **a**, **b**, and **c**, then the angle between bonds **a** and **c** when looking directly along bond **b** is the torsional angle. It is most easily seen by drawing a Newman projection.

The two torsional angles in the backbone between each peptide bond are designated φ and ψ (called **phi** and **psi**).

When φ and ψ are both 0°, the backbone is completely planar. As φ and ψ are rotated, so different conformations are produced, although some combinations are forbidden because of steric interactions.

$$\varphi = 0°$$
φ increases as shown

$$\psi = 0°$$
ψ increases as shown

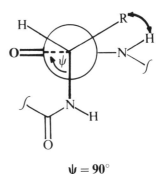

φ = 90°
Steric repulsions are small
Conformation with $\varphi \simeq 90°$ are common

ψ = 90°
Steric repulsions (⌢) are large
Conformations with $\psi \simeq 90°$ are rare

By studying the non-bonding interactions, it is possible to produce the so-called **Ramachandran** plot. This is a two-dimensional representation of φ and ψ, in which contours (or colours) show the likelihood of certain combinations of φ and ψ. In the diagram below, φ/ψ with no adverse non-bonding interactions are shaded, those with minor steric crowding are dotted, and those with major steric interactions are white. As can be seen, only a small range of φ/ψ combinations are 'allowed'.

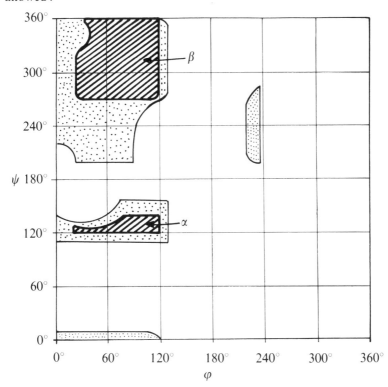

Ramachandran plot. This is an energy contour plot for the φ/ψ combinations about a peptide bond. A typical Ramachandran plot for a dipeptide of two L-amino acids is shown on page 185; low energy shaded regions are high-probability φ/ψ combinations, dotted regions are less probable (medium energy), whilst the white areas indicate high-energy conformations that are sterically disallowed.

Based purely on steric interactions between the α-NH, the C=O, and the β-C, the Ramachandran plot shows that only a limited number of conformations are possible. Nevertheless, if we take 10° jumps in the values of φ and ψ, and if we reject 'disallowed' conformations from the Ramachandran plot (i.e. ignore all the white regions), there are still about 200 φ/ψ combinations allowed for a dipeptide. With the same constraints, there are about 8,000,000 φ/ψ combinations allowed for a tetrapeptide. And that is ignoring the conformations that the side-chains could adopt.

However, it turns out that there are only a small number of important **secondary structures** for which the residues have 'allowed' φ/ψ values. These are the so-called α and β structures, and they correspond to low-energy regions on the Ramachandran plot; once formed, these relatively stable conformations are held together by **hydrogen bonds** between the amide groups:

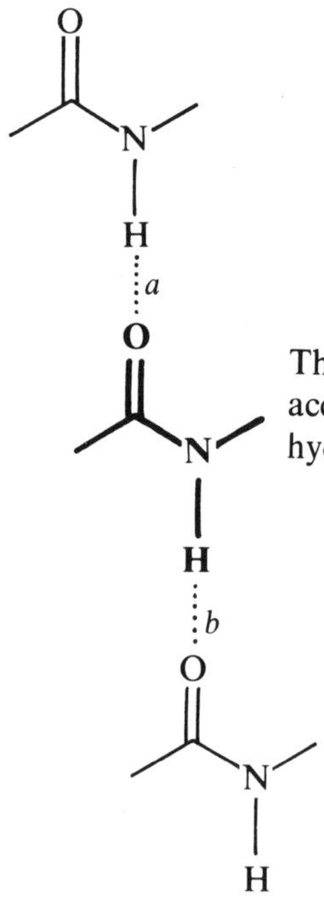

The C=O of an amide can act as a hydrogen-bond acceptor (a). The N—H of an amide can act as a hydrogen-bond donor (b).

2   Secondary Structures

When there is no particular conformation adopted by a stretch of residues, this is termed **random coil**. But there are also well-defined secondary structures, which are stabilised by specific patterns of hydrogen bonding. These structures fall into three groups.

2.1   Helices

Most proteins and some peptides contain helical stretches, in which the corkscrew structure is held together by a regular series of hydrogen bonds.

For example, every C=O involved in a particular helix might be bonded to every **fourth** NH group, as shown below:

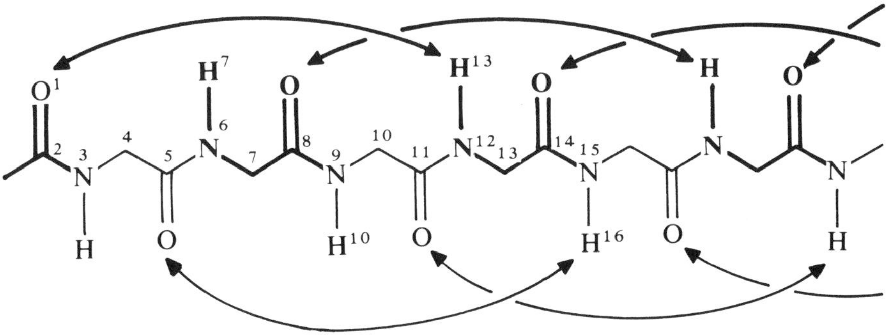

This is called a 3.6_{13} helix, because there are 3.6 residues per turn of helix, and the hydrogen bond is between O(1) and H(13). Helices can be formed by repeated H-bonding between O(1) and H(7), or H(10), or H(13), or H(16), etc. This leads to **right-handed helices**. Interestingly, H-bonding in the other direction is not possible within a helical structure, because the C=O and NH groups cannot be satisfactorily orientated for **repeated** H-bonding (but loops and turns are possible—see later).

The 3.6_{13} helix is particularly important because it is (usually) the most stable of the helices.

> The 3.6_{13} helix is called the **α-helix**. Its stability is due to the relatively unstrained H-bonds, and to the fact that the bulky side-chains are kept well apart.

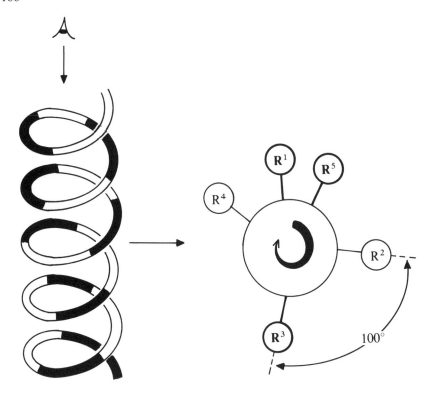

The **α-helix**. The α-helix has 3.6 residues per turn in a right-handed helix. The side-chains are kept well apart, with 100° between adjacent R-groups.

Whilst helices are common in proteins (particularly the α-helix), peptides rarely have sufficient **length** for the H-bonds to really stabilise the structure.

No. of residues	< 5	5	6	7	8	20	40
No. of H-bonds	0	1	2	3	4	16	36
H-bonds/residue	0	0.20	0.33	0.43	0.50	0.80	0.90

H-bonds per residue for α-helices

The Table shows that short α-helices have only a small amount of H-bond stabilisation per residue; other conformations are usually preferred due to other factors (e.g. H-bonds with solvent, or entropic effects). Nevertheless, some important naturally occurring peptides do possess regions of α-helix; and certain synthetic peptides that have high α-helix forming potential (e.g. poly-alanine) have been used to study α-helix formation and stability.

2.2 Sheets

The two secondary structures that dominate the conformation of **proteins** are the **α-helix** and the **β-pleated sheet**.

Like the α-helix, the β-sheet conformation is maintained by H-bonds. But whilst the α-helix requires a coiled backbone, in the β-sheet it is almost fully extended, and involves H-bonding between two distinct stretches of protein or peptide. The β-sheet can take one of two forms.

2.2.1 *Anti-parallel β-pleated sheet.* In this case, the two stretches of peptide or protein are orientated in opposite directions, as shown below:

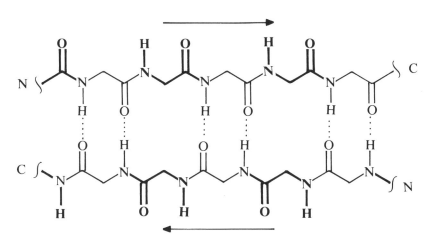

H-Bonding in anti-parallel β-pleated sheet

Although it appears that the peptide chain is fully extended (in this two-dimensional representation), the chains are in fact puckered (or '**pleated**'), with the side-chains extending **away** from the sheet to reduce steric crowding.

2.2.2 *Parallel β-pleated sheet.* This structure is also puckered, and involves stretches of protein that are oriented in the same direction. As with the anti-parallel sheet, it is the 'pleated' conformation that reduces steric crowding of the side-chains.

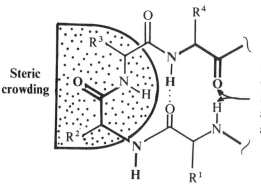

H-Bonding in parallel β-pleated sheet

With anti-parallel and parallel β-sheets, there is an average of one hydrogen bond per pair of residues; i.e., if there were 10 residues in each strand of a β-sheet, then there would be a total of 10 H-bonds between the strands. Therefore, from an energetic point of view, β-sheets are as viable for peptides as for proteins.

In actual fact, the **parallel β-sheet** does **not** occur in **peptides**, because there are never enough residues to allow the chain to loop round—but it does occur frequently in proteins. In contrast, the **anti-parallel β-sheet** occurs in a number of cyclic peptides, in which the peptide backbone makes a hairpin bend.

2.3 Turns and Loops

If a peptide or protein has a relatively compact structure, then it is necessary for the backbone to fold back on itself. There are a number of conformations that allow this to occur, the most important of which is the β-turn.

Steric crowding

A second H-bond can often occur, helping to further stabilise the β-turn. This can also mark the start of an anti-parallel β-pleated sheet.

The 4 → 1 loop or β-turn

It involves a hydrogen bond between the C=O of one residue and the N—H three residues **back** in the sequence (cf. helices), leading to a 10-atom loop. The region around the loop is sterically crowded; this crowding can be reduced, and the ease of β-turn formation improved, in three ways:

(i) removing the R^2/R^3 side-chains (i.e. Gly residues);
(ii) inverting the stereochemistry of R^2 or R^3 (i.e. D-residues);
(iii) cyclising R^2 or R^3 onto the adjacent NH (i.e. Pro residues).

In cyclic peptides, **two** β-turns are usually required, and the residues involved are almost always those that facilitate β-turn formation.

3 Methods of Structure Determination

There are many ways of studying the structure of peptides. The most detail can be obtained from X-ray crystallography or NMR spectroscopy, but other methods can yield valuable information about conformation.

3.1 X-ray Crystallography

In theory, when X-rays (or any beams of electromagnetic radiation) are directed at a sample that is composed of uniformly spaced layers, a diffraction pattern can be obtained, in accordance with the Bragg equation. In the case of X-ray diffraction by a crystalline solid, the layers correspond to the uniform spacing of the crystal lattice; the diffraction pattern therefore contains information about the size and shape of the unit cell. [The X-ray diffraction (or scattering) is actually caused by the electrons (or electron density) in the sample.]

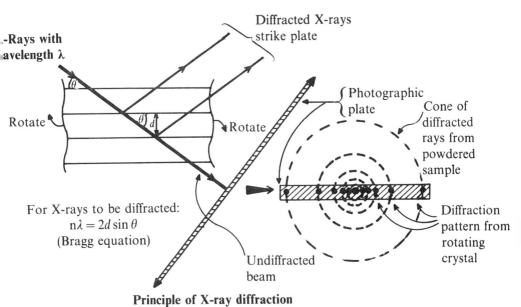

Principle of X-ray diffraction

Bragg equation. For X-rays to be diffracted:

$$n\lambda = 2d\sin\theta$$

$$\text{Path difference} = 2d\sin\theta$$

Diffraction occurs if the path difference $= n\lambda$. The spacing between the layers (d) can be inferred by varying θ, and observing when spots appear on the photographic plate. For many simple ionic compounds, a powdered crystalline solid can effectively generate the full range of θ values, and a pattern of rings (powder pattern) can be observed. But single crystals are required in order to determine the covalent structure of most organic molecules.

For a single crystal, the full diffraction pattern is produced if the sample is rotated within the X-ray beam. When the beam is at certain angles (the **Bragg diffraction angle**) reflection from adjacent (or indeed any two) layers can lead to constructive interference (i.e. the emerging X-rays are **in phase**) and diffraction occurs. At all other values of θ, reflection leads to destructive interference, and the beam instead passes straight through the sample.

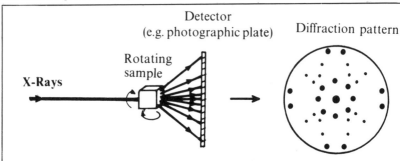

Diffraction pattern. If the crystal is rotated in the X-ray beam, then a 2-dimensional array of spots is observed, which is indicative of the 3-dimensional structure of the sample.

The **pattern** of dots from such a single crystal X-ray diffraction experiment only tells you the shape and size of the **unit cell** (i.e. three lengths, a, b, and c and three angles, α, β, and γ); in other words, it is the unit cell that dictates the spacing and angle of the 'layers' that the X-ray beam 'sees'. However, the **intensity** of the diffracted X-rays is influenced by the electron density within the unit cells.

So, by analysis of X-ray diffraction data, it is possible to construct an electron density map of the unit cell (by a process called **Fourier transformation**), from which the atomic positions can be inferred. For complex molecules like proteins (and peptides), this can be a very difficult job, requiring an enormous amount of computer processing. Moreover, the diffraction pattern records only the **intensity** of the emerging X-rays; phase information (i.e. the sign of the diffracted rays) is lost, and it transpires that structures cannot be solved without it. For proteins, phasing information is usually obtained by the incorporation of heavy atoms into the crystal (isomorphous replacement) before re-collecting the data and processing it. For peptides, direct methods are often successful (where likely phases are added at the Fourier transformation stage). But the 'phase problem' is one of the main obstacles to the rapid solution of X-ray diffraction patterns.

The X-ray crystal structure determination of peptides and proteins can take months (or even years) to carry out, even if suitable crystals can be prepared. In terms of molecular detail, X-ray diffraction methods of structure determination are unbeatable.

For most peptides, however, the method has two drawbacks:

(i) Many peptides cannot be crystallised, because their molecular shape is so poorly defined (i.e. they're too floppy).
(ii) If successful, crystallisation might yield samples suitable for X-ray diffraction—but there is always doubt that this is the biologically active conformation; the crystallisation process might have trapped the molecules in a regular but biologically inactive conformation.

For the peptide chemist, X-ray crystallography can be an extremely valuable tool. But even if crystallisation of a peptide is successful, the structural results must be treated with a certain amount of caution.

3.2 Nuclear Magnetic Resonance (NMR)

NMR is emerging as one of the most powerful modern methods of studying the structure of peptides and proteins. Unlike X-ray crystallography, the compound is studied in solution; this means that structural results from NMR are usually a good indication of the conformations adopted by the molecule in its natural state.

NMR spectroscopy probes the environment of certain nuclei—for work on peptides (and proteins) these nuclei are usually ^1H and ^{13}C, both of which have a spin of $\frac{1}{2}$.

Spin $\frac{1}{2}$ nuclei. The two nuclear spin states are equivalent (or degenerate), under normal circumstances. But if a magnetic field (B_0) is applied, two energy levels are created, with most nuclei being in the lower energy state. If the nuclei are irradiated with radiation of the correct energy (ΔE), then absorption takes place as nuclei are excited from the lower to the higher energy states. For typical magnetic fields ($ca.\ 2.5\,T$), this corresponds to the frequency (v) of radio waves at around 10–600 MHz (depending on the nucleus and the applied field). Because the two states are very close in energy, there is only a small population difference (about 1 nucleus in 10^6). Consequently, high-quality NMR spectra require relatively large amounts of sample (i.e. milligrams rather than micrograms).

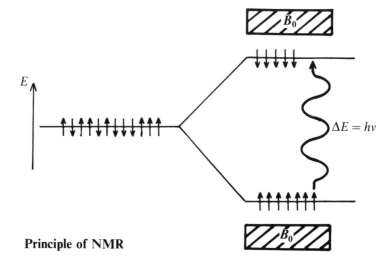

Principle of NMR

NMR spectra record three features of the nuclei being studied:

(i) **chemical shifts**;
(ii) **couplings**;
(iii) **integrations**.

All these features carry important structural information, and we will consider their relevance to a simple dipeptide (Gly—Ala).

^{1}H NMR spectrum of

Gly—Ala in D_2O at 100 MHz

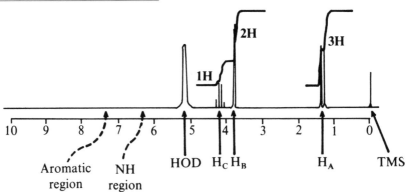

(i) **Chemical shift.** The greater the electron density around a nucleus, the smaller will be the effect of the magnetic field (B_0), and the smaller the chemical shift. On the other hand electron-withdrawing groups increase the chemical shifts of nearby nuclei, by reducing the shielding of the nuclei.

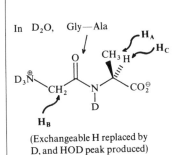

In D_2O, Gly—Ala

(Exchangeable H replaced by D, and HOD peak produced)

Chemical shifts are measured in parts per million (p.p.m. or δ) relative to tetramethylsilane (Me_4Si or TMS) as standard reference. i.e If a proton resonates at 100 Hz higher frequency than TMS (on a 100 MHz NMR spectrometer), then its chemical shift is $\delta 1.0$. (Chemical shift is independent of the field; the same proton would resonate at 200 Hz more than TMS on a 200 MHz spectrometer). But different nuclei resonate at very different frequencies in the same magnetic field:

Nucleus	Natural abundance	Range (δ)	v for TMS at $B_0 = 2.35T$
^1H	100%	0–10	100 MHz
^{13}C	1%	0–200	25 MHz

For example:

H_A:δ 1.35
H_B:δ 3.80 ⎱ Deshielded relative to H_A, by
H_C:δ 4.15 ⎰ adjacent C=O and N

If the pD cf. (pH) is dropped (add $D^{\oplus}$), then H_C is less shielded ($-CO_2^{\ominus} \rightarrow -CO_2H$), and its δ value increases. If the pD is raised (add $\overset{\ominus}{O}D$), then H_B is less deshielded ($-\overset{\oplus}{N}D_3 \rightarrow -ND_2$), and its δ value decreases.

(ii) **Couplings.** Coupling is caused by the 'through bond' interaction of nuclei, and leads to NMR signals being split into multiplets [typically doublets (d), triplets (t), quartets (q), or combinations of these]. Proton signals are usually only split by protons on the adjacent atom (vicinal coupling). For example:

Proton	Multiplicity	Vicinal ^1H(s)	Coupling constant (J)	
H_A	d	$H_c(1)$	7 Hz	$J = 7$ Hz is typical of freely rotating CH groups
H_B	s	None	—	
H_C	q	$H_A(3)$	7 Hz	

Both the multiplicity and the coupling constant (distance between adjacent peaks in a multiplet, measured in Hz) carry important structural information. For ^{13}C spectra, the protons can be fully decoupled from the ^{13}C nuclei ($\rightarrow$ singlets), or partially decoupled (called off-resonance) so that only directly bonded protons split the ^{13}C signals.

Karplus equation. If two vicinal protons are coupled to each other, this can be recognised because they will have identical coupling constants (J). The J-value tells us about the tortional angle between the protons, and is described by the Karplus equations.

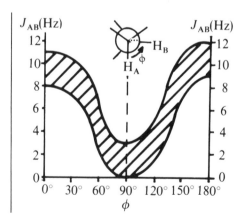

The graph shows the expected range of J_{AB} values for different tortional angles ϕ. For peptides, the coupling between the N—H and the α-C—H protons can be used to determine the angle ϕ (or average angle if the peptide doesn't have a fixed conformation).

(iii) **Integration.** The area under an NMR peak can be measured quite accurately (integration). For protons, the integration is a measure of the number of protons contributing to the signals. For example:

Proton	Integration
H_A	3
H_B	2
H_C	1

For ^{13}C spectra, the integration is not usually a quantitative measure of the number of identical ^{13}C nuclei, due to relaxation effects (the rate at which the population of the energy levels is re-established after irradiation). Moreover, ^{13}C spectra are much weaker than ^1H spectra, because only 1% of carbon nuclei are ^{13}C.

NOE. The nuclear Overhauser effect (NOE) can be used to identify protons that are close in space ($<$ ca. 6 Å). In the peptide below, if H_B was irradiated whilst recording the ^1H NMR spectrum, then the integration of H_A would alter by a few %.

This intensity change is hard to see unless the original spectrum is subtracted—this would produce no net signal for H_C (not close in space to H_B), but a peak at the correct δ for H_A would be observed; the NOE difference spectrum would thereby indicate the proximity of H_A to H_B.

Pulsed NMR and Fourier transformations. NMR spectra can be obtained by scanning across the frequency range whilst monitoring the absorption. But the rapid development of NMR in recent years has been largely due to **pulsed** NMR, in which the sample is irradiated with a pulse of radio waves possessing a range of frequencies; after a short delay (the induction period), the **intensity of radiation emitted** over a period of time is recorded. This **free induction decay** (FID) contains all the information needed to produce a normal NMR spectrum, by carrying out the mathematical process of **Fourier transformation** (FT). High-quality spectra can be obtained by carrying out **data acquisition** from many pulses (typically several thousand). The power of FT-NMR is that **multi-pulse sequences** can be used; these can be used to generate two-dimensional spectra, like the one shown for Gly—Ala below.

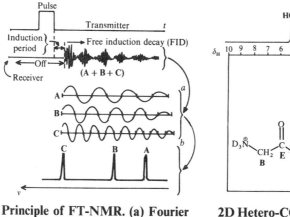

Principle of FT-NMR. (a) Fourier transformation of the FID. (b) Normal NMR representation

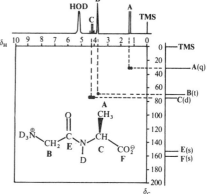

2D Hetero-COSY Spectrum of Gly— Ala in D₂O at '100 MHz'

In the 2D hetero-COSY spectrum of Gly—Ala, only three cross-peaks are observed; these indicate which carbons and hydrogens are directly bonded to each other. The 1D ^{1}H and ^{13}C spectra are shown along the axes, so that the cross-peaks can be assigned; the multiplicities next to the ^{13}C peaks refer to the off-resonance splittings, due to attached protons, and the displayed ^{13}C spectrum is fully ^{1}H decoupled.

For the dipeptide Gly—Ala, the two-dimensional spectrum does not reveal much additional information. But for bigger peptides, they can enable extremely complicated coupling patterns to be unravelled, or complex NOEs to be assigned.

When bond angles (from *J* values), inter-atomic distances (from NOE experiments), other data (e.g. slow deuterium exchange of intra-molecularly H-bonded NH groups in D_2O), and conformational restrictions (e.g. from Ramachandran plots) are all combined, then it is often possible to propose the most likely conformations of a peptide—and sometimes a unique three-dimensional structure can be deduced.

The use of NMR for peptide structure determination is an extremely skilled business. Much expertise is needed simply in order to obtain good spectra—and the analysis of complex spectra can take many months. But the technique is particularly powerful because it can be applied to any peptide (not only those that crystallise—cf. X-ray crystallography), and the structural information applies to the peptide in solution (and so probably resembling its natural environment). Often NMR can show that certain spectral properties are due to a fluctuating structure, thereby helping to define which regions in a peptide have well-defined three-dimensional structure. There can be no doubt that this already important technique will really start to dominate work on peptide conformation in the next few years.

3.3 Other Techniques for Structure Determination

3.3.1 *Optical Methods (ORD and CD)*. Most amino acids are optically active molecules; solutions of them will cause plane-polarised light to be rotated clockwise or anti-clockwise. Secondary structures in peptides have additional optical properties (e.g. the right-handed α-helix) that may increase or reduce the cumulative optical rotation expected from the constituent amino acids. It turns out that simple measurements of optical rotation cannot be used to study these features reliably, but the **variation of optical properties with wavelength of the plane-polarised light** (λ) can reveal features of secondary structure:

Optical rotatory dispersion (ORD) plots optical rotation vs λ.
Circular dichroism (CD) plots absorption of plane-polarised light vs λ.

Provided that the peptide is UV active at the wavelengths studied, both ORD and CD can provide information about secondary structure—especially the content of α-helix.

3.3.2 *Spectroscopic Methods (UV and IR)*. Both of these spectroscopic techniques can reveal some structural information.

Using FT-IR in D_2O, the C=O stretches of the amide groups are characteristic of certain secondary structures. For example, α-helices give peaks at about $1651 \, cm^{-1}$, whilst $1658 \, cm^{-1}$ is typical of disordered structure.

UV spectroscopy is only of value with UV active peptides containing Phe, Trp, or Tyr. It can sometimes be used to reveal unusual environments for these residues (e.g. lipophilic surroundings, or H-bonded tyrosine). Distances between these residues can sometimes be deduced by energy transfer experiments (e.g. irradiate Tyr at its λ_{max}, and monitor the UV emission at λ_{max} for Trp).

3.3.3 *Quantitative Structure Activity Relationships (QSAR).* Many peptides of interest are biologically active, and the biological properties of a series of related peptides can be very informative about structure.

Synthetic analogues

For example, if D-Ala—Ser were found to elicit a particular biological response, then analogues A and B might be prepared and tested. If analogue A stimulated the same response as D-Ala—Ser, then the two peptides would be presumed to adopt the same geometry **in the active conformation**. If analogue B failed to show biological activity, then structure C might be surmised to contain the key structural features for biological activity. This could then be used as the basis for developing new, more effective analogues.

This type of 'trial and error' approach may sound rather unscientific. But it does allow the **structure of the biologically active conformation of a peptide** to be studied directly. It is a tactic that is widely used by medicinal chemists; it is often part of research programmes aimed at developing better drugs (more active or with higher selectivity), and the analysis of analogues sometimes reveals compounds of great pharmaceutical value.

3.3.4 *Structure Prediction.* With all the advances in conformational analysis and structure determination, you might think that chemists ought to be able to predict the three-dimensional structure of small molecules like peptides. But even the backbone of a simple tetrapeptide has 8 million possible conformations (if Ramachandran 'allowed' values of φ and ψ are considered, in 10° increments — see pages 184–186). Worse still, the biologically active conformation of a peptide bound to its target protein might not be the lowest energy conformation when free in solution. Even so, the techniques of molecular modelling are producing some impressive results, especially when used in conjunction with experimental data from structure determination studies (e.g. NMR, QSAR, etc.). There are three main approaches to structure prediction (which may be used on their own or combined).

Statistical analysis. This approach was initially studied by Chou and Fasman, who analysed the frequency with which particular residues occurred within different types of secondary structure. For example, α-helices often contain consecutive tripeptides with polar/non-polar/non-polar side-chains; if a peptide has a sequence that shows this pattern, then an α-helix is likely to be formed. Statistical analysis can also determine the hydrophilicity/lipophilicity of peptides—this tactic is widely used to infer which **sequences** in a **protein** are hydrophilic, and hence likely to be on the outside of the three-dimensional structure; if this hydrophilic peptide sequence is synthesised, it may be used to raise antibodies both to itself, and to the protein from which it was derived.

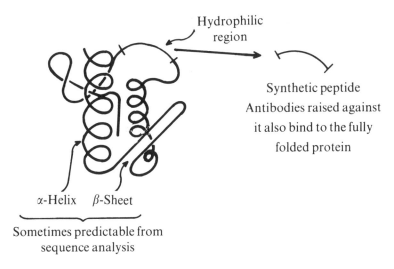

Hydrophilic region

Synthetic peptide
Antibodies raised against
it also bind to the fully
folded protein

α-Helix   β-Sheet

Sometimes predictable from
sequence analysis

Sequence homology. If the structure of a peptide is unknown, but that of a similar related peptide is known, then it is probable that the two three-dimensional structures will be similar. This use of sequence homology is particularly powerful with proteins, especially when used in conjunction with energy minimisation (see below).

Energy minimisation. In theory, the most stable three-dimensional structure for a peptide can be elucidated by calculating which conformation has the lowest energy, using standard values for steric factors/H-bonds/ionic interactions, etc.— but the sheer number of possible conformations precludes a complete analysis of this type. However, by starting with a reasonable proposed structure (e.g. from statistical analysis or sequence homology), it is possible to allow the atoms to 'relax' into their lowest energy state. This does not necessarily give the lowest energy structure—a completely different conformation, accessible only through a high-energy intermediate structure, might actually be preferred.

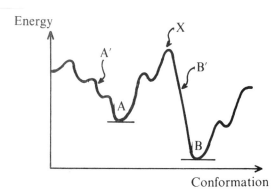

Peptide conformations vs energy. Energy minimisation starting from conformation A' might lead to the *local minimum* A. The *global minimum* B would only be found by passing through the high energy conformation X, or if minimisation was started from structure B'.

Molecular dynamics tries to discover such **global energy minima** by putting energy into the molecule (either overall, or to a specific atom) between steps of energy minimisation. These methods are becoming more successful as our understanding of peptide structure improves, and as our computational power increases.

Peptide structure determination will never be fully perfected. By their very nature, most peptides possess considerable conformational freedom. But by combining the enormous range of techniques that are available for structure analysis, impressive advances are being made to our understanding of the conformations adopted by peptides.

Further Reading

Secondary Structure

The Structure and Action of Proteins, R. E. Dickerson and I. Geis, Harper and Row, 1969. Easy to follow, and really well illustrated.
Principles of Protein Structure, G. E. Schulz and R. H. Schirmer, Springer-Verlag, 1979. Covers many important aspects of secondary structure, but quite hard.

Methods for Structure Determination

NMR in Chemistry: A Multinuclear Approach, W. Kemp, Macmillan, 1986. This is not about NMR of peptides, but it is one of the most friendly introductions to NMR in general.

The Peptides; Analysis, Synthesis, Biology, Volume 7, *Conformation in Biology and Drug Design* (S. Udenfriend and J. Meienhofer, Eds.), Academic Press, 1985. A very useful volume, covering many aspects (both theoretical and practical) of peptide structure; Chapter 9 (by H. Kessler *et al.*, pp. 437–73) gives an excellent indication of the power of NMR in the structural analysis of peptides.

Physical Biochemistry, K.E. van Holde, Prentice-Hall, 1971. Chapter 11 contains a clear, simple summary of the principles of X-ray diffraction.

The Peptides; Analysis, Synthesis, Biology, Volume 4, *Modern Techniques of Conformational, Structural, and Configurational Analysis* (E. Gross and J. Meienhofer, Eds.), Academic Press, 1982. Chapters 1 and 2 (pp. 1–84) discuss X-ray crystallography of peptides, and are particularly useful.

Drug Design

J. G. Vinter, *Chem. Br.*, 1985, **21**, 32, and C. H. Hassall, *Chem. Br.*, 1985, **21**, 39. These two 'Chemistry in Britain' articles make easy reading, and give an indication of the value of computers in drug design.

C. R. Beddell, *Chem. Soc. Rev.*, 1984, **13**, 279. This review article is entitled 'Designing drugs to fit a macromolecular receptor'.

Examples of Structure Determination

D. H. Williams, *Chem. Soc. Rev.*, 1984, **13**, 131. This review, entitled 'Structural studies on bio-active molecules', outlines how the author has used NMR and FAB mass spectrometry to elucidate the structure of complex naturally occurring peptides.

B. Bodo *et al.*, *J. Amer. Chem. Soc.*, 1985, **107**, 6011. A delightful example of structure determination using FAB mass spectrometry, ^{13}C NMR, ^{1}H NMR, 2D NMR (COSY), NOE, amino acid analysis, and chiral g.c.; the structure of Trichorzianine A IIIc, an anti-fungal peptide, was finally elucidated.

Peptide Sequencing and Synthesis Using DNA/RNA Technology

Since the mid-1970s, our ability to understand and manipulate the genetic code has revolutionised the ways in which molecular biology is studied. In particular, the **sequencing** and **synthesis** of proteins has been simplified dramatically, due to the development of techniques that utilise DNA/RNA technology. But for a number of reasons, these methods of **genetic engineering** are rarely suitable for work on peptides, although recent developments may change this in the near future.

In order to see how DNA/RNA technology can be applied to the study of peptides and proteins, we need to understand how their structure is 'hidden' within the genetic code, and how the code directs their **biosynthesis**. Then we can consider how this knowledge can be utilised in order to sequence peptides and proteins, and how the synthesis of large quantities of them can be achieved using genetic engineering.

1 Biosynthesis of Proteins

1.1 The Polymeric Backbone of DNA/RNA

All living cells use exactly the same genetic code, and almost identical pathways in order to generate proteins.

The genetic code is initially carried on **DNA** (deoxyribonucleic acid). In all cells, there is a master copy of DNA (in the form of a double helix), and this directs both protein biosynthesis and cell division (replication).

Protein biosynthesis is not controlled **directly** by the DNA. Instead, a second coded sequence called RNA (ribonucleic acid) is produced from the DNA master copy, and it is the RNA that controls the protein sequence during its biosynthesis. So the sequence of events is as follows:

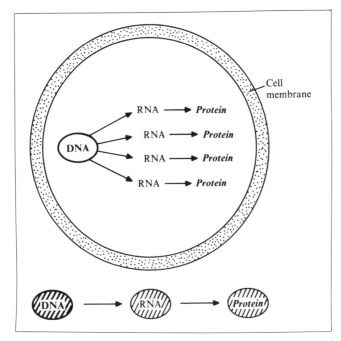

Both DNA and RNA are polymers that are composed of the following repeating unit:

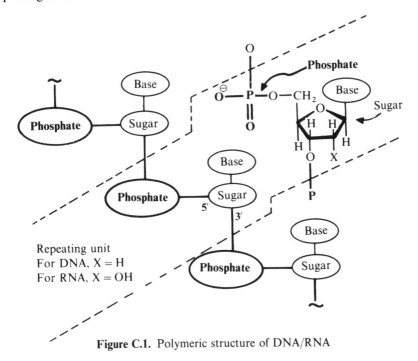

Repeating unit
For DNA, X = H
For RNA, X = OH

Figure C.1. Polymeric structure of DNA/RNA

The **polymeric backbone** in DNA/RNA contains alternate phosphate and ribose units. As in peptides, the **direction** of the polymer must be specified. By convention, DNA/RNA sequences are expressed in the 5' to 3' direction (see Figure C.1).

1.2 DNA/RNA Bases

The genetic code in DNA/RNA is determined by the sequence of **bases** attached to the polymeric backbone. Both DNA and RNA utilise **four** bases, which have the following structures, names, and one-letter abbreviations:

The four DNA/RNA bases

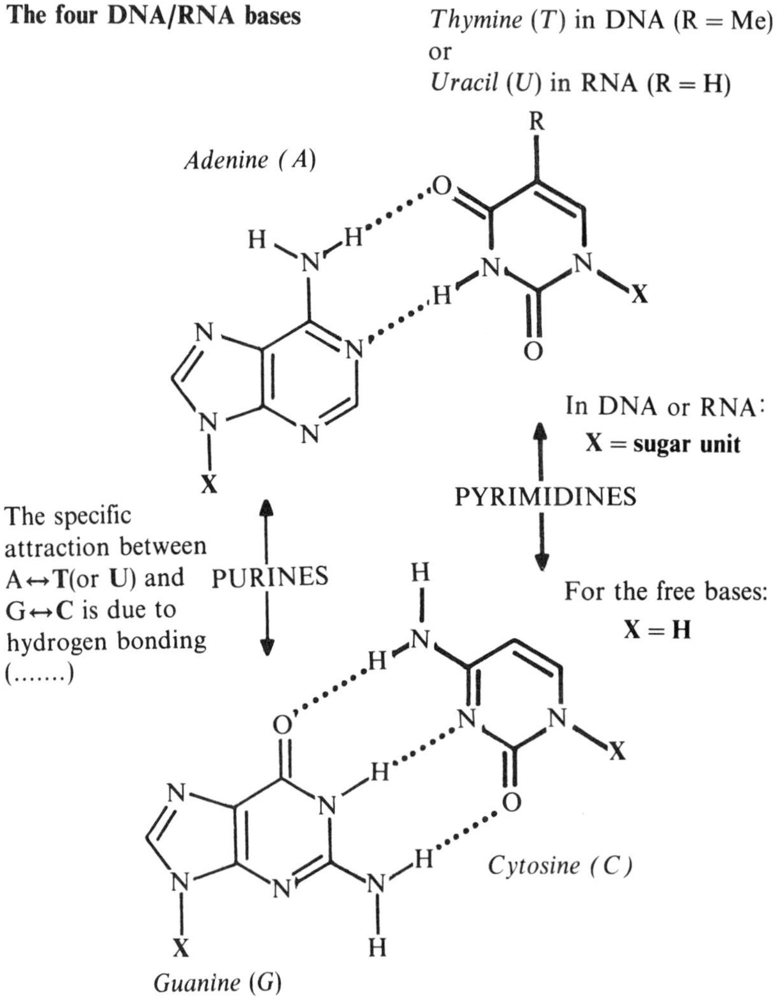

Thymine (T) in DNA (R = Me)
or
Uracil (U) in RNA (R = H)

Adenine (A)

In DNA or RNA:
X = **sugar unit**

PYRIMIDINES

The specific attraction between A↔T(or U) and G↔C is due to hydrogen bonding (.......)

PURINES

For the free bases:
X = H

Cytosine (C)

Guanine (G)

Because all DNA/RNA possesses the same backbone (except for the extra oxygen in RNA) it is only the order of the attached bases that dictates the genetic code (cf. the side-chains defining the sequence of a peptide). So the code can be simplified to the sequence of the attached bases, each of which has a one-letter abbreviation.

For example, **CTA** is a **trimer** of **DNA** (because it contains **thymine** instead of **uracil**); the bases **cytosine**, **thymine**, and **adenine** are attached to the polymeric backbone **in that order** (going from the 5′ to the 3′ end) as shown below:

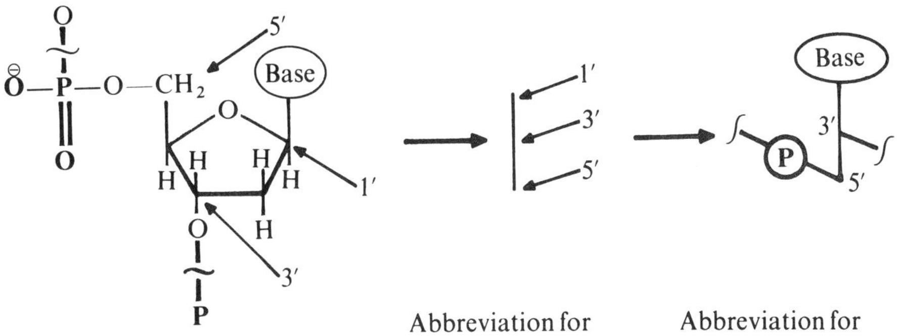

DNA repeating unit

Abbreviation for deoxyribose (sugar)

Abbreviation for DNA repeating unit

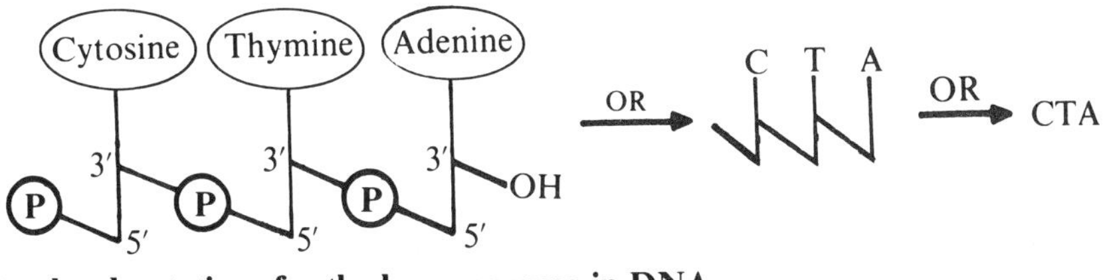

Shorthand notations for the base sequence in DNA

Cells fall into two categories:

Prokaryotes. Simple one-celled organisms (e.g. bacteria). Their DNA floats freely within the cell. Additional pieces of circular DNA (plasmids) can be readily introduced into these cells.

Eukaryotes. More complex multi-celled organisms (e.g. all animals, insects, plants, and fungi). Their DNA is contained within the nucleus of the cell, and is less easy to manipulate.

Bases, nucleosides, and nucleotides. The DNA/RNA bases are all purines or pyrimidines.

A nucleoside is a base joined on to a sugar (either ribose or deoxyribose).

A nucleotide is a phosphorylated nucleoside; mono-, di-, or tri-phosphates can be formed:

e.g. Adenosine triphosphate (ATP)

1.3 Cell Division

The DNA double helix is held together by hydrogen bonding between the bases on each strand. It is a **specific base pairing** between A↔T and between C↔G (as shown on page 206), and it ensures that the two strands of the DNA are truly complementary; i.e. if the sequence on one strand is known, then that on the other strand can be inferred.

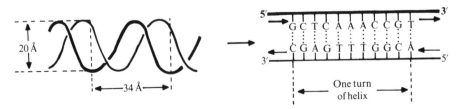

The classical right-handed double helix of DNA. There are 10 bases per turn.

Linear representation of DNA, showing the base pairing.

When cell division takes place (i.e. when one cell divides into two), it is essential that both 'daughter' cells contain a complete and perfect copy of the original DNA. This can take place by the two strands of the 'parent' DNA being separated, and then each strand picking up complementary bases during the enzymatic formation of two new double helices.

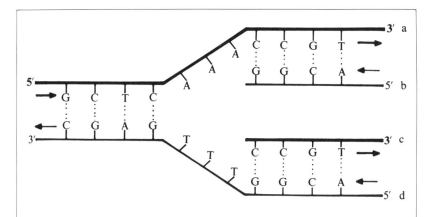

DNA replication. The original 'parent' DNA is a double helix composed of **a** and **d**. When cell division occurs, each strand of the double helix picks up complementary bases, producing two double-stranded identical 'daughters' (**a**↔**b** and **c**↔**d**). A perfect copy of the 'parental' DNA is thereby encoded into the two 'daughter' double helices, which can be incorporated into two cells.

Hybridisation. The complementary strands of DNA are held together by many hydrogen bonds. But if we possessed a solution of just single-stranded DNA, we could test whether another solution contained the complementary strand by seeing if they bound together. In practice, the probe DNA is usually radio-labelled, whilst the DNA to be tested is immobilised onto a plate or filter (see Figure C.2 on page 210).

If the radio-labelled DNA sticks to the plate, then hybridisation has taken place, and the two stretches of DNA are complementary. So reliable is the technique that the temperature of the experiment can be chosen so that only a perfect match will produce a positive result. Hybridisation is widely used to check the integrity of genetically engineered DNA.

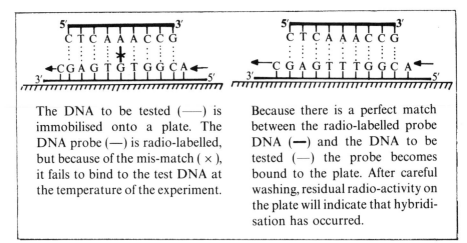

The DNA to be tested (——) is immobilised onto a plate. The DNA probe (—) is radio-labelled, but because of the mis-match (×), it fails to bind to the test DNA at the temperature of the experiment.

Because there is a perfect match between the radio-labelled probe DNA (—) and the DNA to be tested (——) the probe becomes bound to the plate. After careful washing, residual radio-activity on the plate will indicate that hybridisation has occurred.

Figure C.2. Testing DNA matching by hybridisation

1.4 Formation of RNA

Exactly the same type of base pairing is used in the formation of RNA from DNA, so that $A \leftrightarrow U$ and $C \leftrightarrow G$. When a cell needs to **express** a particular genetic sequence (i.e. produce the protein encoded by a particular DNA sequence), then the DNA strands are separated in the appropriate region, and a complementary RNA copy of one of the strands is made.

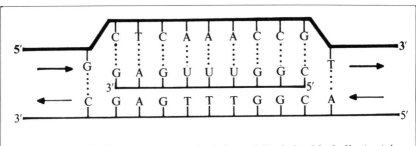

Formation of RNA. One strand of the original double helix (—) is used as a template for the formation of complementary RNA (—).

These RNA polymers act as templates for the biosynthesis of peptides and proteins, whose amino acid sequences are governed by the genetic code.

1.5   The Genetic Code

The genetic code within RNA is contained in **triplet codons**, i.e. the base sequences are **read** in **threes**, with each triplet designating a particular amino acid (plus start/stop commands). The formation of peptides and proteins from RNA uses ribosomes—enzymes that interact with successive triplet codons, and then attach the next appropriate amino acid to the growing peptide or protein. The genetic code is shown in Table C.1 on page 212.

Modern genetic engineering relies on two crucial features of protein biosynthesis:

(i) If the DNA or RNA sequence is known, then the corresponding protein sequence can be inferred. This has greatly simplified protein sequencing.
(ii) If a cell contains a DNA sequence that encodes for a specific protein, then (given the correct stimuli) the cell is capable of producing large quantities of that particular protein. This has revolutionised the synthesis of proteins.

2   Peptide and Protein Sequencing

If it is possible to determine the base sequence in DNA, then the primary structure of the encoded peptides or proteins can be inferred. But is DNA sequencing any easier than amino acid sequencing? The answer is an emphatic 'yes', and two extremely powerful methods have been developed; both of them rely on the generation of labelled DNA fragments.

2.1   The Maxam and Gilbert Method

This approach uses chemical means to generate specific DNA fragments, and it is particularly suited to the study of short stretches of DNA (e.g. those that might encode for a peptide).

Suppose we started with the following DNA nonamer (which might have been isolated or synthesised):

CTCAAACCG

Firstly, a readily identifiable tag is introduced onto one end of the DNA; this is usually a radio-label (^{32}P phosphate or ^{35}S thiophosphate) which replaces the 5' phosphate by an enzymic reaction.

CTCAAACCG $\longrightarrow$ *P-CTCAAACCG

Next, the labelled DNA is divided into four portions, each of which will receive different chemical treatment. This treatment comes in two stages, involving

Table C.1. The triplet codons of the genetic code

1st Base (5' end)	2nd Base				3rd Base (3' end)
	A	G	C	U	
A	Lys	Arg	Thr	Ile	A
	Lys	Arg	Thr	Met	G
	Asn	Ser	Thr	Ile	C
	Asn	Ser	Thr	Ile	U
G	Glu	Gly	Ala	Val	A
	Glu	Gly	Ala	Val	G
	Asp	Gly	Ala	Val	C
	Asp	Gly	Ala	Val	U
C	Gln	Arg	Pro	Leu	A
	Gln	Arg	Pro	Leu	G
	His	Arg	Pro	Leu	C
	His	Arg	Pro	Leu	U
U	**STOP**	**STOP**	Ser	Leu	A
	STOP	Trp	Ser	Leu	G
	Tyr	Cys	Ser	Phe	C
	Tyr	Cys	Ser	Phe	U

cleavage of the base from the backbone, followed by hydrolytic removal of the resulting base-free sugar:

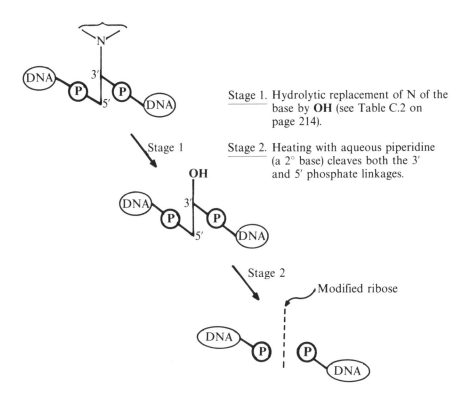

Stage 1. Hydrolytic replacement of N of the base by **OH** (see Table C.2 on page 214).

Stage 2. Heating with aqueous piperidine (a 2° base) cleaves both the 3′ and 5′ phosphate linkages.

For sequencing purposes, only sufficient reagent is added at stage 1 to destroy (on average) about one base per DNA molecule. For example, if cleavage at **cytosine** was carried out, the following **labelled** DNA fragments (oligonucleotides) would be formed from the nonamer on page 211 (plus a complex mixture of unlabelled fragments):

*P-CTCAAACCG
*P-CTCAAAC
*P-CTCAAA
*P-CT

The fragments can be readily separated using gel electrophoresis; the smaller the DNA fragment, the further it travels on the gel. The resolution is so good that fragments that differ in length by only one nucleotide can be separated easily.

The radio-labelled oligonucleotides can be visualised by placing the gel on top of a photographic plate or sheet; the plate will be darkened where it is adjacent to radioactive bands—a process known as autoradiography.

Table C.2. Stage 1 conditions for selective base cleavage

	Base(s)	Conditions	Reason for selectivity
Purines	G	Me$_2$SO$_4$ then H$_2$O/Heat	Dimethyl sulphate methylates purines. Products ($>$N$^\oplus$—Me) readily hydrolysed. But G methylates faster than A.
	A + G	Me$_2$SO$_4$ then HCl(aq)	G methylates faster than A. With HCl(aq), [A—Me]$^\oplus$ hydrolyses faster then [G—Me]$^\oplus$. Overall hydrolysis of G $\simeq$ A.
Pyrimidines	C + T	H$_2$N—NH$_2$(aq)	Hydrazine converts pyrimidines into readily hydrolysed diazirines (R—⟨$^{H}_{N}$NH)
	C	H$_2$N—NH$_2$/NaCl(aq)	The presence of 2M-NaCl suppresses reaction with thymine.

So, after our radio-labelled DNA nonamer has been partially fragmented under the four sets of base-specific conditions, a simple gel electrophoresis would yield the following autoradiogram:

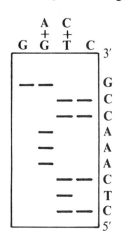

Autoradiogram ladder of CTCAAACCG, after selective base cleavage using the Maxam and Gilbert method, and fragment separation by electrophoresis.

Amazingly, the base sequence can be simply read off from the pattern of bars. Only one problem remains to assigning the amino acid sequence—determining which strand of the DNA has been sequenced. This is because the genes in cells consist of two complementary sequences of DNA, only one of which is the template for RNA formation and protein biosynthesis.

> Stretches of DNA that carry out some biological function (e.g. encode for a protein) are called **genes**.
> The entire DNA of a cell is known as its **genome**.
> The genetic code of DNA and RNA is contained in triplet **codons**, with each codon representing a specific amino acid (or start/stop signals). Only one strand of the DNA (the **template**) is transcribed into RNA; the other strand is said to contain **anti-codons.**

If our DNA nonamer is the template strand, then the corresponding RNA would have the complementary sequence (but with U instead of T).

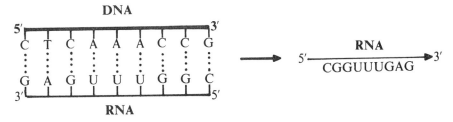

Writing the RNA sequence the normal way round (from 5' to 3'), we can immediately assign the corresponding amino acid sequence from the genetic code.

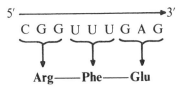

So, in a single experiment, the amino acid sequence corresponding to our DNA nonamer can be elucidated. In fact, this method of analysis can allow up to 200 bases to be sequenced in one experiment.

But there are a number of problems that are often encountered during DNA sequencing, such as determining whether the template or complementary strand has been analysed. But a more fundamental problem is obtaining the correct stretch of DNA at all.

216

RNA is only generated when a cell requires the synthesis of the corresponding protein. So a cell producing large amounts of a specific peptide or protein (e.g. the hypothalamus gland producing LH-RH) will be rich in the corresponding RNA. This may be isolated relatively easily, and will contain only the base sequence that encodes for that particular peptide or protein.

DNA sequencing often starts from RNA, for a number of reasons:

(i) Cells usually contain only one copy of DNA. But if a cell is producing a particular protein of interest, then it will normally generate many copies of the corresponding RNA.

(ii) Even for the simplest of organisms, the DNA content of the genome is millions of bases long—yet only a tiny stretch will correspond to one particular peptide or protein. So a huge amount of effort would be needed in order to determine the entire DNA sequence; or the portion of interest would need to be identified and cleaved from the genome.

(iii) In mammalian cells, a gene in the DNA rarely encodes a peptide or protein directly. Having been transcribed into RNA, whole sections of the RNA may be excised, before the amino acids are assembled together, based on the modified RNA template.

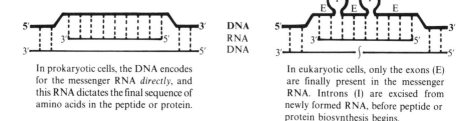

In prokaryotic cells, the DNA encodes for the messenger RNA *directly*, and this RNA dictates the final sequence of amino acids in the peptide or protein.

In eukaryotic cells, only the exons (E) are finally present in the messenger RNA. Introns (I) are excised from newly formed RNA, before peptide or protein biosynthesis begins.

Only if the messenger RNA is sequenced will the correct protein structure be inferred.

For practical reasons, it is usually easier to sequence DNA than RNA. So once the correct RNA has been isolated, a complementary strand of DNA is required, and this can be achieved by using a special viral protein called **reverse transcriptase**. Other microbiological 'tricks' can be used to generate multiple copies of the DNA. It is fair to say that the sequencing experiments themselves are probably the easiest part!

However, there are a number of important reasons why DNA sequencing is not particularly well suited to determining the primary structure of **peptides**:

(i) Many peptides are modified in mammalian cells **after** their initial biosynthesis, e.g. acylation, amidation, or fragmentation of precursor molecules. So DNA sequencing would not generate the correct final structure.

(ii) Many peptides are produced by different biosynthetic pathways, and may contain unusual amino acids, e.g. many fungal antibiotics like penicillin and gramicidin S. So DNA/RNA technology is inapplicable.

(iii) Short stretches of DNA or RNA have other practical problems associated with their isolation and manipulation.

Because of these limitations, direct peptide sequencing will always remain an extremely important method—even though DNA/RNA technology is making astonishing strides forward. The DNA sequencing method of Maxam and Gilbert is particularly important for confirming the sequence of synthetic oligonucleotides (say less than 100 bases)—the importance of these compounds will be discussed later.

Proteins are usually analysed by a different method of DNA sequencing that was developed by Sanger. It is particularly well suited to these larger molecules, which are difficult to analyse using direct sequencing of the constituent amino acids.

2.2 The Sanger Method

Sanger's method of DNA sequencing allows enormous stretches of DNA to be analysed rapidly. It relies on an extremely important enzyme called DNA polymerase.

> **DNA polymerase** is an enzyme that acts on single-stranded DNA. It catalyses the formation of a second **complementary** strand, using nucleotide triphosphates as the source of the new DNA.

If single-stranded DNA is treated with radio-labelled nucleotide triphosphates in the presence of DNA polymerase, then the radio-label will become incorporated into the newly forming complementary strand.

When the positions of the cytosine bases (for example) are to be probed, a small amount of the dideoxycytosine triphosphate is also added to the above reaction:

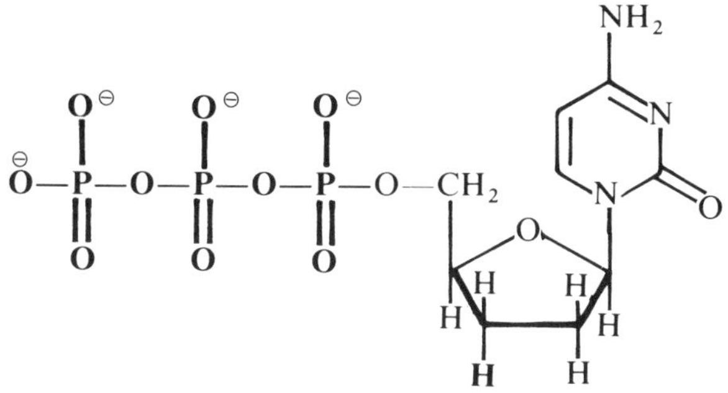

Dideoxycytosine triphosphate

The dideoxy analogue has all of the molecular features necessary for incorporation into the new DNA instead of cytosine, except that it lacks a 3′–OH group, so DNA synthesis cannot continue after its uptake. The amount of dideoxy analogue is carefully chosen to produce the full range of DNA fragments that terminate with cytosine. The DNA fragments can then be separated by electrophoresis; because the normal nucleotides are radio-labelled, the positions of the cytosine bases can be determined by autoradiography of the gel.

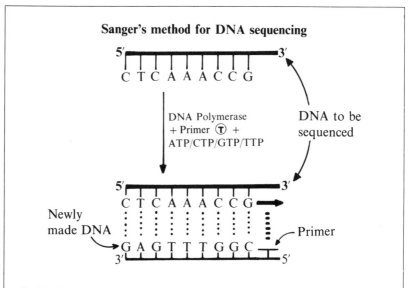

Sanger's method for DNA sequencing

(i) DNA polymerase catalyses the formation of a complementary strand of DNA (5′ → 3′).

(ii) A short *primer* sequence is needed to get the process started—this must be complementary to the 3′-end of the DNA to be sequenced Ⓣ.

(iii) The double-stranded DNA can be separated into single strands by heating.

(iv) The newly formed strand can be identified if the trinucleotide 'food' contains radio-labelled bases.

(v) Dideoxy-bases halt the DNA biosynthesis, giving fragments whose lengths indicate the positions of the corresponding DNA bases.

(vi) The fragments can be separated by electrophoresis.

By carrying out the same procedure with the dideoxy analogues of the other three bases, the full DNA sequence can be determined.

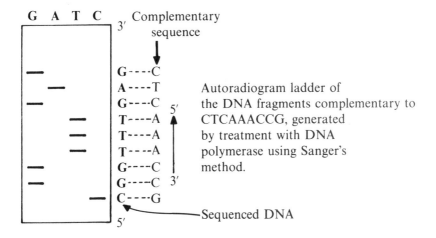

G   A   T   C

Autoradiogram ladder of the DNA fragments complementary to CTCAAACCG, generated by treatment with DNA polymerase using Sanger's method.

Sequenced DNA

As with the Maxam and Gilbert method, one major problem is to obtain the correct DNA to start with—and again, it is usually derived from RNA by treatment with reverse transcriptase. But there are a number of other difficulties with Sanger's method, although they have been largely overcome by some very neat molecular biology.

The DNA processing required before sequencing can begin may seem rather long-winded and cumbersome. But in fact, they are relatively simple and reliable to carry out, although Sanger's method is not well suited to the sequencing of peptides. In contrast, it is an extremely elegant and efficient way of determining the primary structure of proteins, and of sequencing very long stretches of DNA. The technique has undoubtedly spearheaded the spectacular advances in molecular biology over the last 15 years.

3   Peptide and Protein Synthesis

Many of the advances in molecular biology have depended upon the preparation of large quantities (i.e. milligrams) of peptides and proteins. For peptides, chemical assembly of the constituent amino acids is usually the best way—particularly for non-DNA encoded peptides. But DNA/RNA technology can allow us to 'engineer' the DNA in certain cells, in order to make them produce a particular peptide or protein in large amounts. This requires four steps:

(i) obtaining the desired DNA;
(ii) choosing a suitable cell line;
(iii) introducing the DNA into the cell;
(iv) getting the cell to actually produce the peptide or protein.

DNA amplification

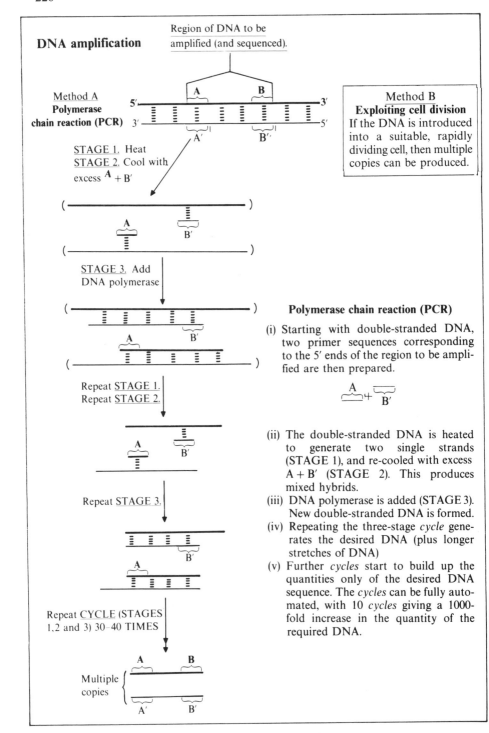

Region of DNA to be amplified (and sequenced).

Method A
Polymerase chain reaction (PCR)

STAGE 1. Heat
STAGE 2. Cool with excess A + B′

STAGE 3. Add DNA polymerase

Repeat STAGE 1.
Repeat STAGE 2.

Repeat STAGE 3.

Repeat CYCLE (STAGES 1,2 and 3) 30–40 TIMES

Multiple copies

Method B
Exploiting cell division
If the DNA is introduced into a suitable, rapidly dividing cell, then multiple copies can be produced.

Polymerase chain reaction (PCR)

(i) Starting with double-stranded DNA, two primer sequences corresponding to the 5′ ends of the region to be amplified are then prepared.

(ii) The double-stranded DNA is heated to generate two single strands (STAGE 1), and re-cooled with excess A + B′ (STAGE 2). This produces mixed hybrids.

(iii) DNA polymerase is added (STAGE 3). New double-stranded DNA is formed.

(iv) Repeating the three-stage *cycle* generates the desired DNA (plus longer stretches of DNA)

(v) Further *cycles* start to build up the quantities only of the desired DNA sequence. The *cycles* can be fully automated, with 10 *cycles* giving a 1000-fold increase in the quantity of the required DNA.

3.1 Obtaining the Desired DNA

There are three ways of achieving this:

(i) Directly from natural sources.

(ii) From RNA using reverse transcriptase. The advantages of this approach were discussed earlier (page 216), and this tactic is widely used (see below).

(iii) Chemical synthesis. This is particularly suitable for peptides, as DNA of up to 100 residues can be synthesised fairly reliably. Solid-phase techniques are usually employed, and excellent protection/activation/deprotection cycles have been developed (cf. peptide synthesis). One widely used combination is shown below:

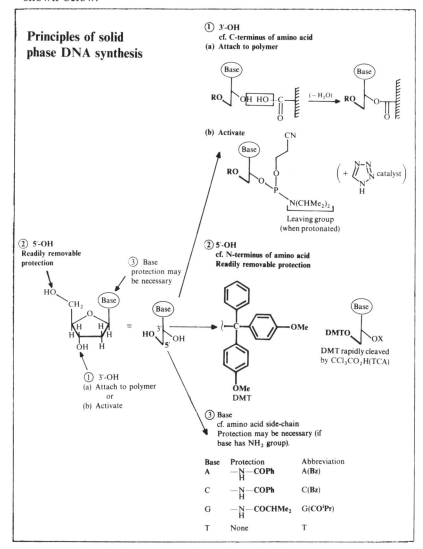

222

Quite often, four resins are kept in stock, each attached to one of the four bases. This means that the cycle of deprotection/coupling/oxidation can be started whenever required, and that fully automated systems can be used. The final deprotection (including removal from the resin) uses $NH_3(aq)$. The application of this methodology to the synthesis of **ACGT** is shown below.

Solid-phase DNA synthesis of ACGT

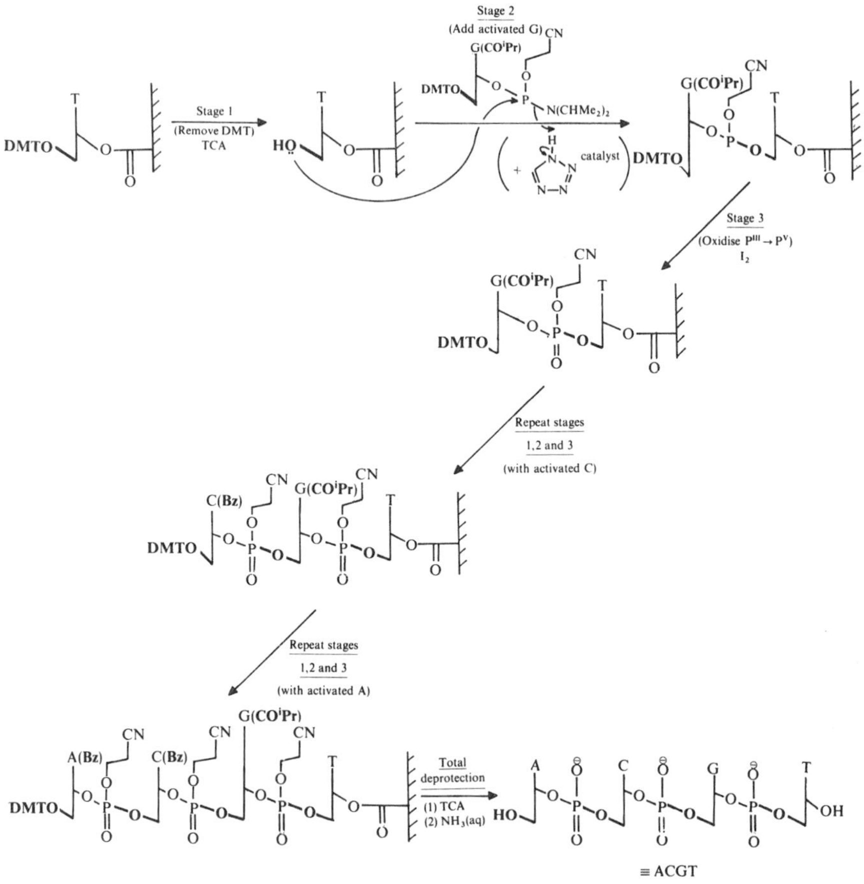

3.2 Choosing the Cell Line

Simple prokaryotic cell lines are usually chosen as the 'host' cell; some of the reasons for this are outlined in the Table below.

Table C.3. Some properties of cells

Property \ Cell	Prokaryote	Eukaryote
(1) Division (i.e. cell growth)	**Easily and rapidly grown in flasks.**	Often hard to culture.
(2) Membrane permeability	**Simple membrane. Readily traversed**	Complex membrane. Hard to traverse.
(3) Location of DNA	**Floats freely within cell.**	Located in nuclear matrix
(4) Simplicity of transcription	**Only *exons* in DNA. ∴ DNA → RNA → Protein**	*Exons and introns* in DNA. ∴ DNA → RNA → Modified RNA → Protein

The most widely used bacterial cells are those of *E. coli*; it is very well studied, easy to handle, and readily available—it lives naturally in the human gut!

3.3 Introduction of DNA

With prokaryotic cells (e.g. bacteria), the introduction of DNA is relatively simple—the membrane can be easily traversed, and the DNA does not have to be incorporated into the nucleus.

Bacterial cells can be made reasonably permeable to certain forms of DNA by placing them in calcium chloride solution. In this state they can readily take up small stretches of circular DNA called **plasmids**. This plasmid DNA is treated by the cell as if it were its own genomic DNA; so, when the cell divides, it makes a copy of both the plasmid and the genomic DNA. The problem is to get the required DNA into a suitable plasmid, and this involves the use of a number of highly specific enzymes:

224

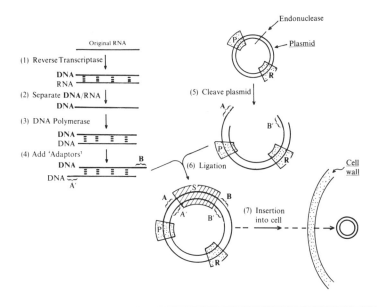

Incorporating DNA into bacterial cells

1, 2. DNA can be prepared from RNA by using reverse transcriptase, then separating the DNA/RNA strands; or it can be prepared directly by chemical synthesis.

3. DNA polymerase generates double-stranded DNA.

4. Adaptor sequences need to be enzymically added, to facilitate uptake of the DNA in step 6.

5. A suitable plasmid is cleaved with an endonuclease enzyme.

6. Mix the two DNA fragments. If the primers have been correctly chosen, they should be complementary to the bases generated by the endonuclease (step 5). Repair enzymes fill in and join any gaps.

7. Add the modified plasmid to the bacterial cells, in the presence of calcium chloride solution to make the walls permeable.

The plasmid in the bacterial cells has three key regions of DNA.

Sequence (S), which is the actual DNA being studied.

Promoter (P), which is a DNA sequence that promotes expression of the sequence (see below).

Resistance (R), which is a DNA sequence that confers cellular resistance to an antibiotic, and allows identification of those cells that have taken up the plasmid (see below).

The resulting cells can be made to divide (producing multiple copies of their DNA and the plasmid), and the promoter can induce the sequence being studied to actually stimulate production of the encoded peptide or protein.

When *E. coli* cells are transfected by plasmids, most of them fail to take up any of the "foreign" DNA. But because the cells can be made to divide rapidly, they can be spread very thinly onto a plate of nutrient, and allowed to form colonies. All of the cells in the colony are identical (clones), and they are all derived from **one** initial cell. The original plasmid is always chosen to have a DNA sequence that makes *E. coli* resistant to a particular antibiotic, which is added to the growth medium. So only those cells that have the plasmid incorporated will be able to grow. If these colonies are then removed from the plate and grown in liquid medium, large amounts of *E. coil* can be produced; and each cell will have a copy of the plasmid, including the DNA sequence of interest. The next requirement is to persuade the cells actually to **produce** the **corresponding peptide or protein**.

3.4 Expression

When a peptide or protein is produced by a cell, the DNA that encodes it is said to be **expressed**. For complex organisms (e.g. animals), subtle control mechanisms are needed to ensure that the correct DNA is expressed at the right time. But even the simplest cells have enormous amounts of DNA that are rarely (or never) expressed.

The expression of a gene is usually controlled by a **promoter sequence**. This is a stretch of DNA that is located **before** the codons for a peptide or protein; when given the correct stimulus, they trigger expression to take place. Typical stimuli are:

change in pH;
change in temperature;
change in concentration of a specific chemical.

So when a plasmid is being engineered, it is important to incorporate a promoter sequence immediately before the DNA of the required peptide or protein. In the scheme on page 224, you can see that this step has been included.

All of the steps needed to prepare a suitable plasmid, and successfully transfect a bacterial cell line, may seen quite complicated and long winded. And one of the biggest problems is identifying a cell which contains exactly the DNA required. But the beauty of this **recombinant DNA technology** is that only one correctly formed cell is required—cell division can then enable millions of copies to be made. Whenever the protein is required, the cells are simply taken out from storage, grown up in a culture medium, and then triggered to produce the desired protein. The cells are then split open (a process called lysis), and the protein extracted and purified.

Recombinant DNA technology offers an additional feature to molecular biologists, once the desired plasmid has been prepared: the facility to produce analogues of the encoded peptide or protein. This uses a technique called **site directed mutagenesis**, in which specific bases in the genetic code are modified. A range of enzymes allow the plasmid to be cleaved at specific points, and synthetic/enzymic methods allow new sequences to be inserted. Of course, non-DNA encoded amino acids cannot be incorporated using this method.

So DNA/RNA technology is extremely versatile and effective for proteins. Peptides are less amenable because they are frequently not encoded directly by DNA, and because the shorter stretches of DNA/RNA cause other practical problems—for the most part, the direct chemical study of peptides is more successful. But genetic engineering is developing so rapidly that its application to peptide chemistry will undoubtedly increase over the next few years.

Further Reading

Biochemistry, 3rd Edition, L. Stryer, 1988. An excellent biochemistry textbook; well laid out and interesting to read throughout—and not too hard. Part V, on 'Genetic Information' (Chapters 27–34, pp. 647–885) is particularly relevant.

Recombinant DNA: A Short Course, J. D. Watson, J. Tooze, and D. T. Kurtz, Scientific American Books, 1983. Co-authored by Watson of 'DNA double helix' fame, this beautifully presented book explains things very clearly.

The Peptides; Analysis, Synthesis, Biology, Volume 5 (E. Gross and J. Meien-hofer, Eds.), Chapter 1, pp. 1–64, Academic Press, 1983. As part of a volume on 'Special Methods in Peptide Synthesis', this helps to put the genetic engineering approach into context.

Genetic Engineering, J. G. Williams and R. K. Patient, IRL Press, 1988. A useful book on this topic.

Laboratory Techniques in Biochemistry and Molecular Biology, Volume 10, '*DNA Sequencing*', J. Hindley (and R. Staden), Elsevier, 1983. Useful practical information about DNA sequencing methods.

Protein Engineering, P.C.E. Moody and A.J. Wilkinson, IRL Press, 1990. This short book (*ca* 90 pages) is part of the 'In Focus' series, and gives a good, concise overview of protein engineering.

Index and Abbreviations

232

The DNA encoded amino acids; structure, abbreviations, and properties

Name	Structure	Abbreviation 3-Letter[a]	Abbreviation 1-Letter[a]	pI	Acidity/ Basicity[b]	Affinity for water[c]	Character of side-chain	Key feature of side-chain	Molecular formula[d]	Molecular weight[d]
Alanine	Me, CO₂H, H, NH₂	Ala	A	6.02	N	O	Alkyl	—R	$C_3H_7NO_2$	89
Arginine	HN, H₂N, N, H, CO₂H, H, NH₂	Arg	R	10.76	B	H	Guanidine	$-N-\overset{NH}{\underset{H}{C}}-NH_2$	$C_6H_{14}N_4O_2$	174
Asparagine	H₂N, O, H, CO₂H, NH₂	Asn	N	5.41	N	H	Amide	—CONH₂	$C_4H_8N_2O_3$	132
Aspartic acid	HO₂C, H, CO₂H, NH₂	Asp	D	2.98	A	H +	Acid	—CO₂H	$C_4H_7NO_4$	133
Cysteine	HS, H, CO₂H, NH₂	Cys	C	5.02	N*	H	Thiol	—SH	$C_3H_7NO_2S$	121
Glutamic acid	HO₂C, H, CO₂H, NH₂	Glu	E	3.22	A	H +	Acid	—CO₂H	$C_5H_9NO_4$	147
Glutamine	H₂N, O, H, CO₂H, NH₂	Gln	Q	5.70	N	H	Amide	—CONH₂	$C_5H_{10}N_2O_3$	146
Glycine	CO₂H, NH₂	Gly	G	5.97	N	O	H	—H	$C_2H_5NO_2$	75
Histidine	H, N, N, CO₂H, H, NH₂	His	H	7.59	N‡	H	Aromatic	—Ar	$C_6H_9N_3O_2$	155
Isoleucine	Me, Me, CO₂H	Ile	I	6.02	N	L+	Alkyl	—R	$C_6H_{13}NO_2$	131